DFG

**The MAK-Collection
for Occupational Health and Safety**

Part II: BAT Value Documentations

Forthcoming Volumes

Greim, H. (ed.)
**The MAK-Collection
for Occupational Health and Safety**
Part I: MAK Value Documentations Volume 21

2005. ISBN 3-527-31134-3 (Wiley-VCH, Weinheim)

Parlar, H. (ed.)
**The MAK-Collection
for Occupational Health and Safety**
Part III: Air Monitoring Methods Volume 9

2005. ISBN 3-527-31138-6 (Wiley-VCH, Weinheim)

Angerer, J. (ed.)
**The MAK-Collection
for Occupational Health and Safety**
Part IV: Biomonitoring Methods Volume 10

2006. ISBN 3-527-31137-8 (Wiley-VCH, Weinheim)

DFG Deutsche Forschungsgemeinschaft

The MAK-Collection for Occupational Health and Safety

Part II: BAT Value Documentations

Volume 4

Edited by Hans Drexler and Helmut Greim

Working Group
Setting of Threshold Limit Values
in Biological Materials
(Head of the Working Group: Hans Drexler)

Commission for the Investigation
of Health Hazards of Chemical Compounds
in the Work Area
(Chairman: Helmut Greim)

WILEY-VCH

WILEY-VCH Verlag GmbH & Co. KgaA

Prof. Dr. med. H. Greim
Senatskommission zur Prüfung
gesundheitsschädlicher Arbeitsstoffe
der Deutschen Forschungsgemeinschaft
Technische Universität München
Institut für Toxikologie
Hohenbachernstr. 15 -17
D-85354 Freising-Weihenstephan

Prof. Dr. med. H. Drexler
Institut und Poliklinik für Arbeits-,
Sozial- und Umweltmedizin
der Universität Erlangen-Nürnberg
Schillerstr. 25 /29
D-91054 Erlangen

All books published by Wiley-VCH are carefully produced. Nevertheless, authors, editors, and publisher do not warrant the information contained in these books, including this book, to be free of errors. Readers are advised to keep in mind that statements, data, illustrations, procedural details or other items may inadvertently be inaccurate.

Translation and Layout: J.A. Handwerker-Sharman

Prior volumes were published under the title "Biological Exposure Values for Occupational Toxicants" (ISSN 0947-2010)

Library of Congress Card No.: applied for
British Library Cataloging-in-Publication Data: A catalogue record for this book is available from the British Library.
Bibliographic information published by Die Deutsche Bibliothek
Die Deutsche Bibliothek lists this publication in the Deutsche Nationalbibliografie; detailed bibliographic data is available in the Internet at http://dnb.ddb.de.

ISSN: 1860-4978
ISBN-10: 3-527-27049-3
ISBN-13: 978-3-527-27049-1

© 2005 WILEY-VCH Verlag GmbH & Co. KGaA, Weinheim

Printed on acid-free paper

All rights reserved (including those of translation into other languages). No part of this book may be reproduced in any form – by photoprinting, microfilm, or any other means – nor transmitted or translated into a machine language without written permission from the publishers. Registered names, trademarks, etc. used in this book, even when not specifically marked as such, are not to be considered unprotected by law.

Printing betz-druck GmbH, Darmstadt
Binding J. Schäffer GmbH i. G., Grünstadt

Printed in the Federal Republic of Germany

Preface

Biological monitoring has become one of the central and most effective instruments in the prevention of damage to the health after exposure to chemical substances at the workplace. It is an integral part of occupational medical health surveillance and has become indispensable in particular for the monitoring of exposure to substances that can be absorbed by the skin. Biological threshold limit values are necessary for the adequate occupational medical evaluation of the analytical results of biological monitoring.

Scientifically justified threshold limit values in biological materials are, to our knowledge, being compiled and published at present by two institutions. One of them is the German Senate Commission for the Investigation of Health Hazards of Chemical Compounds in the Work Area, of the Deutsche Forschungsgemeinschaft. The other organization is a committee of the American Conference of Governmental Industrial Hygienists (ACGIH).

The German biological exposure values for occupational toxicants and carcinogens are specially drawn up by the Working Group "Setting of Threshold Limit Values in Biological Materials". The concept of the German biological tolerance values (BAT values) as well as the exposure equivalents for carcinogenic working materials (EKAs) are well-known and extensively described in Volumes 1 – 3 of the Critical Data Evaluations for BAT and EKA Values.

However, there are numerous hazardous chemicals at the workplace for which neither BAT nor EKA values can be established. This is especially the case if the external and the internal concentrations cannot be correlated, either as a matter of principal or due to the lack of data. In addition, it may be important to consider non-genotoxic end-points of carcinogenic chemicals. Exposure surveillance should be established for sensitizing-agents even though a level of safety cannot be determined. For substances for which the evaluation of a BAT value or an EKA correlation is problematical, a new threshold limit has been established. The Working Group "Setting of Threshold Limit Values in Biological Materials" has called this the "BLW" (Biologischer Leitwert) and the concept is described in Volume 4.

We are very pleased that after a long period the 4th volume of occupational-medical, toxicological documentation for biological exposure values can be presented. It contains documentation for four important BAT values and four addenda for substances which were already presented in Volumes 1 – 3. The particular significance of biological monitoring for occupational exposure to carcinogenic substances is taken into account in seven of the chapters. In four cases no EKA could be evaluated due to the lack of data. For the new BLW value, documentation is included in Volume 4 for four hazardous substances.

In 2005, a new Ordinance for Hazardous Substances at the Workplace was implemented by the German Ministry for Labour and Social Affairs. Changes resulting from that Ordinance have not been considered in the present volume, e.g. withdrawal of Technical Exposure Limits (TRK).

In accordance with the earlier volumes, a review on "Biological Monitoring and Biological Limit Values (BLVs): The Strategy of the European Union" is presented in the special foreword. This review gives a good overview of the activities of the European Union regarding the present state-of-the-art of limit values in the European Union.

At this point, the members of the Working Group should be thanked for their highly qualified work. It is due to their personal engagement, on a voluntary basis, that this collection of documents could be published. The establishment of the documentation for biological exposure values for occupational toxicants and carcinogens demands, in addition to a high degree of expertise, a considerable investment in time.

We would also like to thank those involved with the secretarial office of the Working Group for their editorial work and dealing with the correspondence, and for the preparation carried out for setting biological exposure values. Our translator, Mrs Handwerker-Sharman, deserves particular praise; with understanding and engagement she has greatly contributed to the success of this volume.

Our particular thanks finally goes to the Deutsche Forschungsgemeinschaft for their continued support of the activity of the Working Group.

Erlangen, February 2005

H. Drexler
K.H. Schaller

Contents of Volume 4

Biological Tolerance Values

A. Concepts of Biological Limit Values (BAT Values, EKA and BLW) XI

B. Occupational-Medical Toxicological Documentation of Biological Limit Values (BAT, EKA and BLW) XXIII

Introduction to Biological Monitoring

Biological Monitoring and Biological Limit Values (BLVs): The Strategy of the European Union 3

BAT Value Documentation

Cyclohexane 13

Formic acid methyl ester 23

Hexachlorobenzene, Addendum 31

Lead and its compounds (except lead arsenate, lead chromate and alkyllead compounds) 39

Lead and its compounds, Addendum 79

Lead and its compounds, Addendum 87

Manganese and its inorganic compounds 89

Tetrachloromethane, Addendum 117

Tetrahydrofuran, Addendum 119

Documentation for Carcinogenic Substances
With Biological Exposure Equivalents

Dichloromethane, Addendum 123

Tetrachloroethene, Addendum 127

Trichloroethene, Addendum 131

Documentation for Carcinogenic Substances
Without Biological Exposure Equivalents

Antimony and its inorganic compounds 141

Beryllium and its inorganic compounds 151

1,4-Dichlorobenzene, Addendum 159

Mercury, organic mercury compounds 161

BLW Value Documentation

Arsenic and inorganic arsenic compounds
(with the exception of arsenic hydride and its salts) 171

Cresols (all isomers) 179

Methyl bromide 189

Phenol, Addendum 201

List of Authors and Date of Compilation 207

Working Group "Aufstellung von Grenzwerten in biologischem Material"
(Setting of Threshold Limit Values in Biological Material) 209

Index for Volumes 1–4 213

Biological Tolerance Values
Significance and Use

A. Concepts of Biological Limit Values (BAT Values, EKA and BLW)

The BAT Concept – Biological Tolerance Value for Occupational Exposures (Biologischer Arbeitsstoff-Toleranz-Wert)

Definition

The BAT value ("Biologischer Arbeitsstoff-Toleranz-Wert": biological tolerance value for occupational exposures) is defined as the maximum permissible quantity of a chemical substance or its metabolites or the maximum permissible deviation from the norm of biological parameters induced by these substances in exposed humans. The BAT value is established on the basis of currently available scientific data which indicate that these concentrations generally do not affect the health of the employee adversely, even when they are attained regularly under workplace conditions. As with MAK values, BAT values are established on the assumption that persons are exposed at work for at most 8 hours daily and 40 hours weekly. BAT values established on this basis may also be applied without the use of correction factors to other patterns of working hours.

BAT values can be defined as concentrations or rates of formation or excretion (quantity per unit time). BAT values are conceived as ceiling values for healthy individuals. They are generally established for blood and/or urine and take into account the effects of the substances and an appropriate safety margin, being based on occupational medical and toxicological criteria for the prevention of adverse effects on health.

Prerequisites

By definition, BAT values can be established only for such substances which can be taken up by the body in substantial quantities via the lungs and/or other body surfaces (skin, gastrointestinal tract) during occupational exposure. Another prerequisite for the establishment of a BAT value is that sufficient occupational-medical and toxicological data are available for the substance and that these data are supported by observations in man. The data must have been obtained with reliable methods. For the establishment of new BAT values and the annual review of the list, the submission of suggestions and reports of experience with such substances in man is requested.

Derivation of BAT Values

The derivation of a BAT value can be based on various constellations of scientific data which reveal a quantitative relationship between exposure concentration and body burden and therefore permit the linking of MAK and BAT values. These include
- studies which reveal a direct relationship between concentrations of a substance, metabolite or adduct in biological material (body burden) and adverse effects on health
- studies which reveal a relationship between a biological indicator (effect parameter) and adverse effects on health.

The following considerations of sex-specific factors apply for the establishment of BAT values:
1. The range of the variation in human anatomical and physiological differences which affect the toxicokinetics of a substance is very wide even for a single sex; the ranges for the two sexes overlap.
2. The resulting sex-specific differences in toxicokinetics vary in a range which is insignificant compared with the uncertainty involved in establishing threshold values.
3. Pregnancy can be associated with certain changes in the toxicokinetics of xenobiotics. In practice, however, the effects of these changes are limited, so that for health protection at the workplace it is the effects on the embryo and foetus which are of particular importance.

Documentation

The reasons why a BAT value was established at a particular level are documented by the Commission for the Investigation of Health Hazards of Chemical Compounds in the Work Area in a loose-leaf collection entitled "Arbeitsmedizinisch-toxikologische Begründungen von BAT-Werten". Some of these documents are available in English in the present series. They combine a critical review of the available data with comments on the values for the parameters which have been shown in practice to make a useful contribution to occupational hygiene.

Purpose

In the context of specific occupational-medical check-ups, BAT values are intended to protect employees from impairment of health at work. They provide a basis for deciding whether the amount of a chemical substance taken up by the organism is harmful or not. For substances that can be absorbed through the skin, individual exposures can be determined only by biological monitoring. When using BAT values, the medical exclusion criteria defined by the "Berufsgenossenschaft" (Employers' Liability Insurance Association) in the guidelines for occupational medical check-ups must be observed. BAT values are not suitable for the derivation, by means of fixed conversion factors, of biological threshold values for long-term non-occupational exposures such as from air pollution or contaminants in food.

Correlations between BAT and MAK Values

When a substance is inhaled under steady state conditions in controlled laboratory experiments, the relationship between the BAT and MAK values can be expressed in terms of pharmacokinetic functions. Under workplace conditions, however, it is not necessarily possible to deduce the level of a substance to which a particular person was exposed from its specific biological parameter in that person because a series of other factors in addition to the amount of substance in the air can determine the extent of exposure of the organism. These factors include the level of physical activity (respiratory minute volume), absorption through the skin and individual variations in metabolic or excretory patterns.

It is therefore particularly difficult to evaluate field studies which describe the relationships between internal and external exposure to substances which can be absorbed percutaneously. Experience has shown that studies of such substances frequently yield discrepant results. These discrepancies are attributed to different levels of dermal exposure under the different study conditions. When evaluating such results to determine relationships between MAK and BAT values, priority should be given to studies in which the data suggest that skin absorption played a minor role.

In general for substances with low vapour pressure which are readily absorbed through the skin, there is no correlation between exposure concentration and body burden. For these substances a BAT value can often be established only on the basis of a relationship between body burden and effect.

In addition, the concentrations of substances in the workplace air may vary with time and the biological parameters may not vary to the same extent. Therefore observance of BAT values does not make it unnecessary to monitor the concentrations of substances in the workplace air. This applies especially for local irritants and caustic substances. When evaluating macromolecular adducts of foreign substances it should be borne in mind that the persistance of these adducts can lead to discrepancies between the pattern of external exposure and the behaviour of the biological parameters. Similar considerations apply for all highly cumulative substances such as heavy metals and polyhalogenated hydrocarbons.

In spite of all these interfering factors and the consequent differences in the definitions of MAK and BAT values, the two thresholds are generally based on equivalent effects of substances on the organism. However, for substances for which the MAK value is not established on the basis of systemic effects but because of local irritation of skin and mucous membranes, a BAT value can still be based on "critical toxicity" resulting from systemic exposure. In such exceptional cases where the MAK and BAT values are based on different end points, the two values do not necessarily correspond.

Surveillance

The protection of the health of the individual, which is the reason for establishing BAT values, can be monitored by periodic quantitative determination of the chemical compounds or their metabolites or of biological parameters in biological material. The methods used must be diagnostically specific and sensitive enough for the purpose, acceptable to the employee and practicable for the physician. The sampling time, that is, the measurement strategy, must take into account both the exposure conditions at the workplace and the pharmacokinetics of the substance. As a rule, especially for substances which accumulate in the organism, this may be achieved by taking samples at the end of a working day after an extended period of work (working week).

During exposure to gaseous substances which are metabolized rapidly and for which the blood/air distribution coefficient is larger than 10, it must be taken into account that the concentrations of the substances in blood and tissues are positively correlated with the level of physical activity.

The concentrations of inhaled gaseous substances in blood and tissues of persons working under hyperbaric pressure have been shown to be correlated positively with the pressure. In such cases the observance of the BAT value must be monitored more frequently as the BAT value is attained in such workers at lower exposure concentrations than in persons working at normal pressures.

Whole blood, serum and urine samples are used as assay materials, occasionally and under certain conditions, also samples of alveolar air. Saliva and hair samples are not suitable assay materials for occupational medical biomonitoring.

The analytical methods must yield reliable results and meet the requirements of statistical quality control (TRGS 410 of the regulations for hazardous chemicals at the workplace). In the collection "Analysen in biologischem Material" (available in English in the series "Analyses of Hazardous Substances in Biological Materials"), the Commission's "Analytical Chemistry" group has compiled a series of methods which may be considered reliable for this purpose.

Evaluation of Analytical Data

Like any results of laboratory investigations, toxicological analytical data can only be evaluated given knowledge of the whole situation. As well as the other medical findings, especially
- the dynamics of pathophysiological processes
- the short-term effects of exposure-free periods
- the long-term effects of ageing
- the specific workplace conditions
- intensive physical activity and unusual conditions of atmospheric pressure

and any individual background exposures must be taken into account.

Results from analyses in biological material are subject to medical discretion. Only the physician who is responsible may interpret the results.

BAT values are established on the basis of the results of scientific studies and practical medical experience.

Allergenic Substances

Depending on individual disposition, allergic reactions can be induced by various kinds of substance, more or less rapidly and in differing degrees of severity after sensitization of, for example, the skin or respiratory passages. The observance of BAT values cannot provide a guarantee that such reactions will not occur.

Mixtures of Substances

BAT values apply as a rule for exposure to pure substances. They are not necessarily applicable for persons exposed to preparations containing more than one toxic substance (blends, mixtures, solutions). This is especially true for BAT values requiring determination of the substance itself or its metabolites. For mixtures of components with similar effects, a BAT value based on a biological parameter can be helpful in the assessment of health risks, as long as it provides a measure of critical clinical-functional effects of the components. The Commission makes every effort to define and publish such criteria for the biological effects of interfering components of mixtures.

The EKA Concept – Exposure Equivalent for Carcinogenic Substances (Expositionsäquivalent für krebserzeugende Arbeitsstoffe)

Definition

BAT values are not established for chemical substances which, by their own action or by that of their reactive intermediates or metabolites, are known to cause malignant growths in man or for which there is good evidence of a human cancer risk because at present it is not possible to specify safe levels of such substances in biological materials. Therefore the handling of carcinogens must take place under the conditions described in Section III of the "List of MAK and BAT Values". The determination of carcinogenic substances in biological material is not carried out for the application of a BAT value in the strict sense but rather for the occupational medical detection and quantification of the individual exposure to the substance. Concentrations of the substance or its metabolites in biological material which are higher than those known to correspond to the concentration of the substance in the workplace air are indicative of additional exposure by other routes, usually percutaneous.

For this reason, the Commission investigates the relationships between the concentration of the carcinogen in the workplace air and that of the substance or its metabolites in biological material ("Expositionsäquivalente für krebserzeugende Arbeitsstoffe", EKA: exposure equivalents for carcinogenic substances). From these relationships, the internal exposure which results from uptake of the substance exclusively by inhalation may be determined.

In addition, the considerations expressed in the section "Correlations between MAK and BAT Values" apply to substances which may be absorbed percutaneously.

The BLW Concept (Biologischer Leitwert)

Introduction

Biological monitoring is the analysis of a hazardous chemical, its metabolites or other parameters linked to the substance in the blood or urine of man. It allows an assessment of individual exposure to the chemical or even the effects. In addition to ambient air monitoring it provides an instrument of surveillance and health protection at the workplace.

For practical reasons, it is important that this exposure or its effects can be evaluated by comparing the result of the analysis with a limit value. For this purpose the Deutsche Forschungsgemeinschaft (DFG) has established biological tolerance values for occupational exposures (Biologische Arbeitsstoff-Toleranzwerte, BAT). Ideally, the BAT values are based on the quantitative relationship between the internal dose of the chemical and its effects on health. A historical example is the correlation between the concentration of lead in blood or δ-aminolaevulinic acid in the urine and the effects of the metal on the haematopoietic system. However, due to low exposure levels at modern workplaces, the quantitative relationship between the internal dose and the detrimental effects on health can very often not be determined. In these cases, the BAT value is deduced indirectly from the maximum workplace concentration (Maximale Arbeitsplatz-Konzentration, MAK), which is the limit value in ambient air. This is possible because the MAK value generally guarantees the absence of adverse effects on health. A typical example is the BAT value for toluene in blood and for o-cresol in urine. In this case, it is assumed that observance of the BAT value implies compliance with the MAK value and that adverse effects on health can therefore be ruled out. For carcinogenic substances the DFG has established exposure equivalents (Expositionsäquivalente für krebserzeugende Arbeitsstoffe, EKA). These are not limit values; they merely show the quantitative relationship between the external and internal dose.

Yet there are numerous hazardous chemicals at the workplace for which neither BAT nor EKA values can be established. This is especially the case if the external and the internal concentrations cannot be correlated, either as a matter of principle or due to lack of data. Examples are
- substances that penetrate the skin
- substances that are taken up orally
- tasks in which respiratory protection is used
- tasks that are performed outdoors.

In addition, it may be important to consider non-genotoxic end points of carcinogenic chemicals. Exposure surveillance should be established for sensitizing agents, even though a level of safety cannot be determined. For substances for which the evaluation of a BAT value or an EKA correlation is problematical, a new threshold limit has been created. The working group Setting of Threshold Limit Values in Biological Material has called this the BLW (Biologischer Leitwert) (see Hallier *et al.* 2001).

Definition

The BLW is the amount of a chemical substance or its metabolites or the deviation from the norm of biological parameters induced by the substance in exposed humans which serves as an indicator for necessary protective measures. BLWs are assigned only for hazardous materials for which the establishment of BAT values is not possible (i.e. for carcinogenic substances and suspected carcinogens in the categories 1 to 3 and for non-carcinogens for which the toxicological data are inadequate).

BLW values are generally established on the assumption that persons are exposed at work for at most 8 hours daily and 40 hours weekly during their working lives.

The BLW is based on occupational-medical information as to the effects of handling the hazardous material together with toxicological data. Since observance of the BLW does not exclude a risk of adverse effects on health, it is necessary to extend our knowledge of the relationships between exposure to the substance, the systemic dose and the resulting risks for health. The BLW values are intended to advance this aim by providing a basis for biomonitoring of exposed persons by the physician. By continual improvement of the industrial situation, occupational hygiene and the protective aspects of work planning, concentrations as far as possible below the BLW should be attained.

Requirements

For substances for which there is no BAT value or for which a value cannot be evaluated, a BLW will therefore be established. These are mainly substances
- which are absorbed readily by the skin (e.g. PAHs, glycol ethers)
- for which at present neither relationships between external and internal exposure nor between internal exposure and the effects can be evaluated, but for which there are sufficient data available from biological monitoring at various workplaces. This is the case when, for example, only relatively few analyses of the workplace air were carried out or the results are not representative of the exposure levels. This often occurs with hazardous substances which are difficult to determine, such as dusts, aerosols etc.
- which at the workplace are ingested to a considerable extent (in the form of dust or as substances bound to dust). Occupational-medical practice has shown that the external exposure of the employee is generally underestimated in such cases. This is also because the actual concentration levels at such workplaces are monitored only very inexactly and on the basis of air samples the exposure cannot be reproduced representatively

- with carcinogenic or genotoxic potential (e.g. aromatic amines)
- with allergic effects, if the sensitizing potential cannot be used at present for establishing a BAT value.

The following must be taken into consideration when establishing BLWs:
- toxicological data, in particular for the mechanism of action and toxicokinetics of the substance
- occupational-medical data
- epidemiology
- background exposure
- existing threshold values in air (MAK)
- validated analytical methods must be available for the selected parameters and the requirements for internal and external quality control must be fulfilled.

Below are some examples to illustrate the problem of setting threshold limit values and offer solutions:

Example 1

Studies performed at workplaces with exposure to polycyclic aromatic hydrocarbons (PAHs) have shown that the excretion of hydroxypyrene in the urine differed by two orders of magnitude between individual workers in spite of comparable ambient exposure. This considerable variation is probably caused by uptake of PAHs by the skin. According to the literature, dermal absorption could account for 50 % to 98 % of the total uptake of PAHs at the workplace (Boogaard and van Sittert 1995, Elovaara et al. 1995, Quinlan et al. 1995, van Rooji et al. 1993a, 1993b).

Even if the concentration of the chemical in ambient air is the same, permeation of the skin can lead to very different internal exposures. Peak values of 700 µg hydroxypyrene per litre urine have been measured. Jongeneelen (1992) calculated that a hydroxypyrene excretion of about 6 µg/g creatinine corresponds to an ambient PAH exposure of about 2 µg benzo(a)pyrene per m^3, which according to the author would lead to a relative cancer risk of 1.3. Thus, at some workplaces substantial improvements in occupational hygiene must be made. This illustrates the necessity for a BLW for evaluating the internal PAH dose.

Example 2

Methyl bromide is used as a fumigant for the control of insects, nematodes and fungi. On account of its marked acute neurotoxic properties, which has led to numerous fatal accidents, methyl bromide is applied solely with respiratory protection independent of the surroundings. Yet methyl bromide is also taken up via the skin, leading to exposure of the fumigators. In addition, the masks may not necessarily be impervious. Since the uptake of methyl bromide by inhalation is prevented by respiratory protection, surveillance of the ambient air at the site of fumigation is insufficient. Therefore a biological monitoring procedure was developed on the basis of serum albumin adducts (determination of *S*-methylcysteine) (Müller *et al.* 1994) and tested in an occupational field study (Hallier 1986). (Note: for reasons of analytical practicability, bromide in serum was later chosen as the biomonitoring parameter for methyl bromide). Among the 28 fumigators tested, two individuals with especially high adduct values were conspicuous, whereas the rest of the workers had adduct values which were only slightly (up to 2.5 times) higher than those in unexposed controls. It could be shown that shortcomings in occupational hygiene were responsible for the two high values.

In this example, a conventional limit value such as the BAT is inappropriate, due to the lack of correlation between the external and the internal dose (respiratory protection equipment, skin absorption). Yet a biological monitoring procedure and data on typical blood levels under workplace exposure are available, so that a BLW could be established. Such a value could lead to an improvement in workplace risk prevention and pinpoint shortcomings in occupational hygiene.

Example 3

Another example is arsenic. According to definition, MAK and BAT values cannot be evaluated for the carcinogenic arsenic compounds. In Germany there was a technically based TRK value of 0.1 mg arsenic per m^3 ambient air. According to the EKA, this level in air corresponds to 130 µg/l arsenic in urine at the end of the shift. Arsenic compounds are not only carcinogenic but are also toxic to the skin, the blood vessels and the nervous system. Adverse toxic effects of arsenic have been reported at urinary excretion levels lower than 130 µg/l (Heyman *et al.* 1956, Lagerkvist *et al.* 1986, 1988).
Therefore, a BLW was evaluated to prevent toxic effects.

Example 4

Sensitization and the induction of allergic symptoms are both concentration-dependent phenomena. Therefore exposure to allergens should be limited. As long as threshold values cannot be derived, sensitization and allergic reactions are not suitable end points for MAK or BAT values. A number of occupational allergens can however be surveilled by biological monitoring (e.g. platinum excretion after occupational exposure to platinum salts) in order to identify hazardous workplaces and to implement occupational hygiene (Drexler and Göen 2000, Drexler *et al.* 1999).

References

Boogaard PJ, van Sittert NJ (1995) Urinary 1-hydroxypyrene as biomarker of exposure to polycyclic aromatic hydrocarbons in workers in petrochemical industries: baseline values and dermal uptake. Sci Total Environ 163: 203–209

Drexler H, Göen T (2000) Der Beitrag des biologischen Monitorings zur Risikoerfassung beim Umgang mit sensibilisierenden Arbeitsstoffen am Beispiel der Diacarbonsäureanhydride. Arbeitsmed Sozialmed Umweltmed 35: 136–145

Drexler H, Schaller KH, Nielsen J, Weber A, Weihrauch M, Welinder H, Skerfving S (1999) Efficacy of hygienic measures in workers sensitised to acid anhydrides and the influence of selection bias on the results. Occup Environ Med 56: 145–151

Elovaara E, Heikkilä P, Pyy L, Mutanen P, Riihimäki V (1995): Significance of dermal and respiratory uptake in creosote workers: exposure to polycyclic aromatic hydrocarbons and urinary excretion of 1-hydroxypyrene. Occup Environ Med 52: 196–203

Hallier E (1986) Arbeitsmedizinische Untersuchungen zur Problematik der Durchführung von Begasungen mit Methylbromid. Deutsche Hochschulschriften Nr. 1089, Verlag Hänsel-Hohenhausen, Egelsbach, 1986

Hallier E, Angerer J, Drexler H, Filser JG, Lewalter J, Stork J (2001) Biologische Leitwerte (BLW). Arbeitsmed Sozialmed Umweltmed 36: 6–9

Heyman A, Pfeiffer JB, Willett RW, Taylor HM (1956) Peripheral neuropathy caused by arsenical intoxication: A study of 41 cases with observation on the effect of BAL. N Engl J Med 254: 401–409

Jongeneelen FJ (1992) Biological exposure limit for occupational exposure to coal tar pitch volatiles at cokeovens. Int Arch Occup Environ Health 63: 511–516

Lagerkvist B, Linderholm H, Nordberg GF (1986) Vasospastic tendency and Raynaud's phenomenon in smelter workers exposed to arsenic. Environ Res 39: 465–474

Lagerkvist B, Linderholm H, Nordberg GF (1988) Arsenic and Raynaud's phenomenon. Vasospastic tendency and excretion of arsenic in smelter workers before and after the summer vacation. Int Arch Occup Environ Health 60: 361–364

Müller AMF, Hallier E, Westphal G, Schröder KR, Bolt HM (1994) Determination of methylated globin and albumin for biomonitoring of exposure to methylating agents using HPLC with precolumn fluorescent derivatization. Anal Bioanal Chem 350: 712-7-15

Quinlan R, Kowalczyk G, Gardiner K, Calvert IA, Hale K, Walton ST (1995) Polycyclic aromatic hydrocarbon exposure in coal liquefaction workers: the value of urinary 1-hydroxypyrene excretion in the development of occupational hygiene control strategies. Ann Occup Hyg 39: 329–346

van Rooji JGM, Bodelier-Bade MM, Jongeneelen FJ (1993a) Estimation of individual dermal and respiratory uptake of polycyclic aromatic hydrocarbons in 12 coke oven workers. Br J Ind Med 50: 623–632

van Rooji JGM, van Lieshout EMA, Bodelier-Bade MM, Jongeneelen FJ (1993b) Effect of the reduction of skin contamination on the internal dose of creosote workers, exposed to polycyclic aromatic hydrocarbons. Scan J Work Environ Health 19: 200–207

B. Occupational-Medical Toxicological Documentation of BAT, EKA and BLW Values

Preliminary Remarks

In the prevention of damage to health caused by substances at the workplace, the evaluation of biological parameters is becoming increasingly important within the framework of occupational health surveillance. Medical strategy is based on a principle which has already proved its worth in other fields of application in occupational medicine: the stress-strain (exposure-effect) concept.

Exposure-Effect Concept

As in other areas of occupational medicine also with toxicological questions the concepts of "Belastung" and "Beanspruchung" corresponding to the English concepts of "stress" and "strain" are used. (In the context of this documentation of BAT values the concepts "Belastung" and "Beanspruchung" have generally been translated by "exposure" and "effect", respectively). Under "Belastung" any influence (dose) is understood which can cause a reaction in the human organism. Changes that are caused by this "Belastung" are, however, described as "Beanspruchung". Individual qualitative and quantitative differences can arise in the strain/effect for a concrete stress/exposure so that the reactions of the organism caused by the stress/exposure receive an individual colouring resulting from the different qualities and capabilities of the person.

The concept can best be illustrated with an example from the area of technology with the rod bending test in the testing of materials. Here it is clear that the bending (strain) of the rod to be tested can only then be predicted from a defined force (stress) working on the rod if the material constant is known. Formulated more generally, this means that the effect on the individual as a function of a given exposure can only be understood if the intervening variables (e.g. functional reserve of an organic system) are known.

The exposure-effect concept is therefore medically of particular importance as it takes into account the congenital or acquired interindividual biological variability and thus places the individual at the centre of attention. In the nature of things, the effect on the person can only be objectified by medical examination.

External Exposure and Internal Exposure

Under workplace conditions the human organism is typically exposed to substances through inhalation. To measure and describe the concentration of the substance in the ambient air of the workplace as the "external exposure" and to limit it by the application of MAK values is an important and indispensable measure of primary prevention. It is, however, much more difficult to quantify other sorts of individual "external exposure" (cutaneous or oral contact with substances at the workplace).

If a substance is absorbed in an inhalational, percutaneous and/or oral way by the body, depending upon its metabolism and kinetics in the organism, a certain concentration of the substance results, the "internal exposure". This quantity, independent of the mode of absorption, is not only a reflection of the total "external exposure"; additionally the intervening variables for the absorption of a substance, such as e.g. the size of the respiratory minute volumes, can influence this quantity. As the "internal exposure" can offer valuable information on the concentration of the substances at the biological site of effect, it is an important parameter of secondary prevention and individual health protection. In many cases the concentration of substances in blood or urine samples can be representative of "internal exposure".

Frequently, however, a biological effect cannot be attributed immediately to the substance but to its metabolites. In these cases the metabolite concentration in biological material must be taken as the parameter of "internal exposure". Thus, the capability of the organism to form and detoxify toxic substances is characterized in such situations as an additional variable intervening between exposure and biological effect.

Effect

The effect resulting from the "internal exposure" can only be deduced from functional changes in the organism. It must, however, be borne in mind that not every deviation of a biochemical parameter from its norm can be immediately regarded as evidence of disease. Changes seem tolerable which even long-term
1. with regular exposure do not result in a disturbance in function or capability to compensate for the effects of exposure,
2. are reversible after the end of exposure,
3. do not intensify the sensitivity of the organism to other external influences, in particular, physical, chemical and microbiological ones,
4. do not endanger the progeny.

The Commission wishes to reveal with its "arbeitsmedizinisch-toxikologische Begründung von BAT-Werten" ("Occupational Medical Toxicological Documentation of BAT Values") which observations, experiences and considerations were decisive for the establishment of a BAT value taking the exposure-effect concept as a basis. At the same time it wishes to highlight with the documentation gaps in the data and provide impulses for further research.

The treatise on each substance is subdivided into several main sections.

Front page

Chemical substance
BAT value/EKA or BLW correlation
Sampling time
Date of evaluation
Synonyms
CAS number
Formula (sums and structure)
Molecular weight
Melting point
Boiling point
Vapour pressure at 20°C
Solubility in water
MAK/TRK [last established]

As an *introduction*, a few additional chemical-physical constants which are relevant to exposure and metabolism may be taken into consideration: physical state of the substance at the workplace [gas, vapour, aerosol, mist or dust], additional exposure to isomers, contaminants etc., information on whether the substance is still used industrially or for how long it has been used or produced.

1. Pharmacokinetics

1.1 Absorption

Inhalational, dermal, gastrointestinal—main route of absorption

Inhalational: Pulmonary absorption rate during, at the end of and after the shift. Parameters which influence absorption: e.g. solubility or particle size, physical activity, body weight.

Dermal: Information on the rate of absorption [$mg/cm^2/h$], determined *in vivo/in vitro*. What role does dermal absorption play in the total absorption?

Gastrointestinal: Information on the rate of absorption and the possibility of occupational and non-occupational oral exposure (contamination of the hands, clothing, food, drinks).

1.2 Elimination

Role of metabolism and storage in elimination; extent and speed, routes of excretion (amount in expired air, urine, faeces)

Elimination half-life of the substance and important metabolites. Signs of accumulation of the substance or of metabolites during one day, a week or years.

1.3 Metabolism

Diagram of metabolism in which the main metabolic pathways as well as proven and postulated metabolites are shown.

Information on the dose dependency of metabolism. At which concentrations are signs of saturation and thus a deviation from first order kinetics to be expected?

Information on known interactions, e.g. with alcohol, drugs, food components.

– Pharmacokinetics (half-life)
– Liability to interference (exogenous and endogenous factors)
– Influence of illness and pregnancy

2. Critical Toxicity

In this section it is explained which toxic effects were decisive for the establishment of the BAT value, whereby often a cross-reference is possible to the loose-leaf collection "Toxikologisch-arbeitsmedizinische Begründung von MAK-Werten" (in English translation in the series "Occupational Toxicants", obtainable from Wiley-VCH, Weinheim, Germany).

3. Exposure and Effects

In a summary of the relevant literature it is shown which relationships exist between external and internal exposure as well as between internal exposure and effects.

3.1 Relationship between external and internal exposure

Short description of the relevant data from

a) laboratory studies on volunteers
b) field studies on occupationally-exposed persons
c) simulation studies—pharmacokinetic modelling

3.2 Relationship between internal exposure and effects

Short description of the relevant data for the internal exposure and sensitive and specific reactions (see Critical Toxicity) from:

a) laboratory studies on volunteers
b) field studies on occupationally-exposed persons
c) simulation studies—pharmacokinetic modelling

4. Selection of the Indicators

The increase in individual safety which BAT values are intended to achieve can only be attained if the methods used for preventive examinations are characterized by as high a diagnostic specificity and sensitivity as possible. At the same time the sampling method should be acceptable for the employee and practicable for the attending physician. These criteria and the available knowledge of critical toxicity, as well as the relationship between exposure and effect are taken into account in the selection of the biological indicators. These can be either parameters of internal exposure or of effect. Blood and urine are the preferred examination materials; in certain cases also exhaled air, secretions, body appendages and biopsy material can be considered. The selection of the indicators is substance-specific.

a) Description of all parameters which come into question

Advantages and disadvantages (specificity, measurability, practicability, influence of exogenous and endogenous parameters)

b) Selected parameter(s)

Information on kinetics (immediate increase after exposure, increase in concentration during exposure, decrease or further increase after the end of exposure, decrease in the concentration of the parameter after the end of exposure, type of decay curve and biological half-life, accumulation during the working week, months or years, main storage organs).

5. Methods

Is the use of particular methods of analysis required for monitoring internal exposure using the suggested BAT/EKA/BLW value?

Which methods are suggested by the working group "Analyses of Hazardous Substances in Biological Materials"?

Which methods of analysis have proven reliable and meet the standards of statistical quality control?

Recommendations for sampling, transport and storage. Indication of the possibility of contamination.

6. Background Exposures

Information on the reference values ("normal values") found with the suggested methods. Indication of possible sources of background exposure from the environment or lifestyle (e.g. food, smoking, drugs, implantations).

7. Evaluation of BAT/EKA/BLW Values

Summary for the selection of the exposure/effect parameter(s) and the setting of sampling times and the value. Medical evaluation of the data according to clinical experience.

8. Interpretation of Data

A summary of the clinical experience so far should simplify the difficult task of medically assessing the data from the overall clinical situation.
    Information on the importance of adhering to the sampling time
    Indication of possible errors in sampling, storage and transport
    Indication of analytical problems (use of a particular method, necessity of hydrolysis)

Information on the possibility of an influence on the values measured by the exposure (percutaneous absorption, non-occupational exposure, interindividual and intraindividual fluctuations due to endogenous or other background levels, co-exposure, significant influence of smoking, drugs, physical activity)

Indication of influences due to the individual (influence of illness, pregnancy)

9. References

Important statements should be attested by accessible quotations from the original literature. There is no claim to completeness, suitable reviews are permissible.

Introduction to Biological Monitoring

Biological Monitoring and Biological Limit Values (BLVs): The Strategy of the European Union

For several decades occupational standards for chemical compounds in the ambient air at workplaces have been developed as a preventive tool in occupational hygiene and medicine. The early scientific roots date back to Lehmann (1886), who was the first to perform systematic studies on workers exposed to some occupational toxicants. By 1938, occupational exposure standards had been established in Germany for about 100 substances (Henschler 1985). In the United States, a provisional list of 8 compounds was compiled in 1942, and the first official list of Threshold Limit Values (TLVs) was issued in 1946 (Cook 1985).

In 1953, Oettel tried to establish a unified European list of standards, but failed (Henschler 1985). As a consequence, most European countries started to use the American TLV list. In 1958, Germany started to develop its own list of MAK values, which was later used, either as a whole or in part, also by some neighbouring countries (Austria, Switzerland, the Netherlands, Sweden). However, differences in the industrial, social and constitutional structure of European countries prevented the establishment of a unified European list of occupational standards (Henschler 1985).

Biological Exposure Values – Initial Developments

Progress in analytical chemistry, toxicokinetics and toxicodynamics provided new ways of monitoring toxicants in biological media. The first official reference to biological exposure limits was made in the preface of the 1974 edition of the TLV booklet. It was noted that measurements may be made of substances to which the worker is exposed, based on analysis of blood, urine, hair, nails, body tissues, fluids and exhaled breath, or the amount of metabolites in tissues and fluids (Cook 1985). In Germany, a first set of three biological limit values (lead, toluene, trichloroethylene) was introduced with the MAK list in 1981 ("Biological Tolerance Values", BAT), and another 6 compounds were added in 1982 (e.g. cadmium, mercury, dichloromethane, halothane, perchloroethylene). More recently, a comprehensive compendium of the strategy of biological monitoring in general was issued (Lehnert and Schaller 2000), and future perspectives were highlighted (Deutsche Forschungsgemeinschaft 2000).

The first American Biological Exposure Indices (BEIs) were published by ACGIH in 1984. At that time, a broad international discussion on biological standard setting was initiated, as seen in contributions from Belgium, Denmark, Italy, Poland, Sweden, Switzerland and the USA at an ACGIH/WHO symposium in April 1985 (ACGIH 1985).

The present list of BEI values contains 41 entries (ACGIH 2004), and the German BAT value list has evaluated 78 compounds (Deutsche Forschungsgemeinschaft 2004), highlighting the considerable potential of biological monitoring strategies.

Development of Occupational Exposure Limits in the European Union

With the growing integration of the European Union (EU), there was a need to harmonize national workplace standards, backed by the EU policy to remove trade barriers between its member states (Thier and Bolt 2001).

In particular, the Council Directive 80/1107/EEC (European Union 1980), amended by Council Directive 88/642/EEC (European Union 1988), introduced the objective of establishing Occupational Exposure Limits. This Directive considered two types of Occupational Exposure Limits, *Binding Limit Values* (BLVs) and *Indicative Limit Values* (ILVs). The latter were intended to be the more common type of limit, reflecting evaluations based on scientific data.

The first set of ILVs was introduced by Commission Directive 91/322/EEC (European Union 1991); these ILVs for 27 chemicals (or groups of chemicals) were proposed by the Commission and agreed by Member States on the basis of pre-existing national positions. At the same time, the Commission assembled an advisory group of experts in the disciplines of toxicology, epidemiology, occupational medicine, occupational hygiene and analytical chemistry. The major role of this "Scientific Committee on Occupational Exposure Limits" (SCOEL, formerly called the "Scientific Expert Group", SEG) was to examine appropriate scientific documentation, usually in the form of Criteria Documents, on the toxicological and other relevant properties of chemicals and to recommend to the Commission values for substance-specific Occupational Exposure Limits ("OELs"; SCOEL 1998). In 2000, the European Union issued a list of 62 chemical substances with Occupational Exposure Limits.

At that time, it was recognised that biological monitoring—entailing the measurement of substances, metabolites or adducts in biological media, or the measurement of non-adverse biological effects induced by the substance—was an important means of protecting the health of workers. Details of the procedures used in deriving both OELs (analysis of substances in ambient workplace air) and Biological Limit Values ("BLVs", use of biological media) were communicated (SCOEL 1999). A key compound with respect to general issues of biological monitoring and Biological Limit Values is lead, for which a BLV of 30 µg/100 ml blood has been proposed (SCOEL 2003).

The time sequence of steps implemented by the European Union is shown in Figure 1.

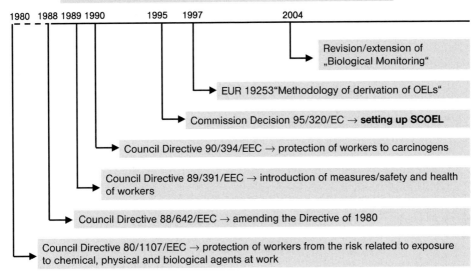

Figure 1. Time sequence of steps towards an official implementation of biological monitoring by the European Union

Skin Notation and Consequences for Biological Monitoring

Skin penetrating compounds may have systemic toxic effects which cannot be efficiently prevented on the basis of analytical monitoring of the workplace air alone. Lists of Occupational Exposure Limits include designations of such compounds ("S" for *skin*, "H" for *Haut* or *huid*, etc.). In Germany, for such compounds (marked "H") the use of biological monitoring strategies has been officially recommended (Anonymous 1996), and an official Technical Guidance Note issued by the Federal Ministry of Labour and Social Affairs describes the legal and organisational framework for biological monitoring (Anonymous 2000).

Biological Monitoring – The Strategy of the European Union

Scientific advances in the biological monitoring of chemical exposures have led to progress in establishing legal frameworks (Bolt 1999).

The European Commission Directive 95/320/EC of July 12th, 1995, which set up the aforementioned Scientific Committee for Occupational Exposure Limits (SCOEL), describes the tasks of the Committee as follows [article 2(1) of this Directive]: "The Committee shall in particular give advice on the setting of Occupational Exposure Limits (OELs) based on scientific data and, where appropriate, shall propose values which may include:
- the eight-hour time-weighted average (TWA),
- short-term limits/excursion limits (STEL),
- biological limit values"

(European Union 1995).

In 1999, SCOEL published the Methodology for the Derivation of Occupational Exposure Limits. This included a first definition of Biological Limit Values, that was later amended: "Biological monitoring entails the measurement of substances and/or metabolites in biological media, and the measurement of biological effects induced by the substance. Biological Limit Values (BLVs) are reference values for the evaluation of potential health risks in the practice of industrial hygiene. They are established on the basis of currently available scientific data. Exposure concentrations equivalent to the BLV generally do not affect the health of the employee adversely, when they are attained regularly under workplace conditions (8 hrs/day, 5 days/week). In general, OELs and BLVs are based on similar quantities of internal exposure; in this case, the BLV is related to a group means. In cases of a high health impact of individual peak exposures, a BLV may be conceived as a ceiling value for individual persons, dependent on its justification."

In addition, details of a biological monitoring strategy have been put forward, as outlined below.

Approaches to Biological Monitoring

Biological methods used to assess exposure and/or risks to health are considered to fall into the following categories:
1. Determination of the substance or its metabolite in a biological medium (*biological exposure monitoring*).

Most methods fall into this category, with the medium of choice usually being blood, urine, or occasionally, exhaled breath. The method may either be specific for a particular substance or general for a group of related substances. The level determined may reflect

exposure over widely different time periods, depending on the kinetics of the substance, the medium involved and the time of sampling.
2. Measurement of biological effects (*biological effect monitoring*).
This involves the measurement of parameters of biological response (e.g. serum cholinesterase activity for organophosphates).

An effective means of biological exposure monitoring is the determination of macromolecular adducts of toxicants or their metabolites (e.g. haemoglobin or serum albumin adducts). The determination of haemoglobin adducts provides an integrated measure of the effective internal exposure over a longer period of time, due to the life span of erythrocytes.

Airborne Limits (OELs) and Biological Limits (BLVs)

BLVs are representative of the levels of determinants which are observed in workers exposed to the respective chemical, exclusively by inhalation, at the level of the respective OELs.

Exceptions are the BLVs for substances for which the OELs serve as protection against non-systemic effects (e.g. irritation or respiratory disorders) or for substances which require biological monitoring due to other routes of absorption, in particular the skin.

BLVs do not indicate a sharp distinction between hazardous and non-hazardous exposures. Due to biological variability, it is possible for an individual's measurement to exceed the BLV without there being an increased health risk. If, however, measurements in specimens obtained from a worker on different occasions persistently exceed the BLV, or if the majority of measurements of specimens obtained from a group of workers at the same workplace exceed the BLV, the cause of the excessive values must be investigated and proper action taken to reduce the exposure.

BLVs for working schedules other than eight hours exposure, five days a week, may be extrapolated on a toxicokinetic and toxicodynamic basis. Attention should be paid to the correct sampling time. In the case of a long half-life of toxicants (e.g. lead), or of biological effect monitoring, the sampling times for biological monitoring are even independent of exposure profiles (as exemplified by Bolt and Rutenfranz 1988).

Biological Media

The choice of biological medium depends on kinetic factors, the convenience of sample collection and the possibility of sample contamination.
- *Blood.* Since this is the main vehicle for transport and distribution, most systematically active substances and their metabolites can be found in blood. The

medium is useful for inorganic chemicals and for organic chemicals which are poorly metabolized and have a sufficiently long half-life.
- *Urine*. Urine collection is easier, less invasive and more readily accepted by workers. It is usually suitable for water-soluble metabolites of organic substances and for some inorganic substances.
The concentration of a substance in urine usually reflects the mean plasma level during the period of urine accumulation in the bladder. End of shift samples are appropriate for rapidly excreted substances, such as solvents; 24-hour specimens (although rarely collected) may be more representative in some cases. The concentration of a substance in urine will depend on the rate of urine production, and correction of results on the basis of creatinine concentration or density may be necessary. Contamination during collection can be a source of error.
- *Breath*. Exhaled air analysis may be used to estimate exposure to volatile organic substances (solvents), although it is much less frequently used than blood or urine sampling. The method is non-invasive, but involves a risk of external contamination.
Concentrations will vary depending on whether they are measured in "end exhaled" air (alveolar) or in "mixed exhaled" air (normal breathing). The timing of sampling is very critical in determining whether the measurement reflects very recent exposure or exposure during the previous day. Concentrations can also be influenced by a variety of physiological factors.

Whichever medium is chosen, it is important to establish a sampling strategy, based on knowledge of the kinetics of the biological marker in question.

Analysis and Interpretation of Results

Careful attention must be paid to both pre-analytical (sample collection, transport and storage) and analytical methodology to ensure accuracy.

Each aspect of biological monitoring should be conducted within an effective quality assurance (QA) programme. The appropriate specimen must be collected, at the proper time, without contamination or loss, and with use of a suitable container. Donor identification, time of exposure, source of exposure and the sampling time must be recorded. The analytical method used by the laboratory must have the accuracy, sensitivity and specificity needed to produce results consistent with the BLV. Appropriate quality control specimens should be included in the analysis, and the laboratory must follow routine quality control rules. The laboratory should participate in an external quality control programme.

Like any results of laboratory investigations, biomonitoring results can only be evaluated given knowledge of the whole situation. As well as the other medical findings, especially
- the dynamics of pathophysiological processes,
- the short-term effects of exposure-free periods,
- the long term effects of ageing,

- the specific workplace conditions,
- intensive physical activity and unusual conditions of atmospheric pressure, and
- any individual background exposures

must be taken into account.

Like any other clinical laboratory data, biological monitoring data need to be interpreted by a physician.

Some BLVs, i.e. those referring to urinary excretion, are expressed relative to creatinine concentrations. In the first instance this refers to compounds for which the relevant studies are documented only on the basis of urinary creatinine values.

Summary and Current Situation

- By decision of 12th July, 1995, the European Commission established the Scientific Committee on Occupational Exposure Limits (SCOEL), with the mandate to give advice on Occupational Exposure Limits (OELs) and Biological Limit Values (BLVs).
- SCOEL has proposed a BLV for inorganic lead in blood of 30 µg Pb/100 ml, and there are currently discussions on a number of compounds to be assigned with a BLV.
- A general strategy related to health-based BLVs has been formulated.
- Compounds with a "skin notation" are viewed as priority candidates for a BLV assignment.
- In general, there is a clear political tendency in Europe away from national and towards supranational (EU) regulations in the entire field of occupational safety and health.
- Improvement of the regulations on chemical substances is receiving high political priority and interest, as exemplified by current discussions on the "White Paper" of the European Commission.

References

ACGIH (1985) International symposium on occupational exposure limits. Ann Am Conf Ind Hyg 12: 1–389

ACGIH (2004) 2004 TLVs and BEIs, based on the documentation of the Threshold Limit Values for Chemical Substances and Physical Agents and Biological Exposure Indices. American Conference of Governmental Industrial Hygienists, 1330 Kemper Meadow Dr., Cincinnati, OH 45240-1643

Anonymous (1996) TRGS 150 Unmittelbarer Hautkontakt mit Gefahrstoffen. Bundesarbeitsblatt 6/1996: 31–33

Anonymous (2000) TRGS 710 "Biomonitoring". Bundesarbeitsblatt 2/2000: 60–62

Bolt HM (1999) Biologische Arbeitsstofftoleranzwerte (Biomonitoring). Teil XIV: Entwicklungen in der Europäischen Union. Arbeitsmed Sozialmed Umweltmed 34: 519–520

Bolt HM, Rutenfranz J (1988) The impact of aspects of time and duration of exposure on toxicokinetics and toxicodynamics of workplace chemicals. in: Notten WRF, Herber RFM, Hunter WJ, Monster AC, Zielhuis RL (Eds), Health surveillance of individual workers exposed to chemical agents. Int Arch Occup Environ Health, Suppl, Springer, Berlin, pp. 113–120

Cook WA (1985) History of ACGIH TLVs. Ann Am Conf Ind Hyg 12: 3–9

Deutsche Forschungsgemeinschaft (2000) Biological Monitoring – Heutige und künftige Möglichkeiten in der Arbeits- und Umweltmedizin, J Angerer (Ed.), Wiley-VCH, Weinheim

DFG, Deutsche Forschungsgemeinschaft (2004) List of MAK and BAT values. Commission for the Investigation of Health Hazards of Chemical Compounds in the Work Area, Report No. 40. Wiley-VCH, Weinheim

European Union (1980) Council Directive 80/1107/EEC. OJ L 327, 3.12.1980, p. 8

European Union (1988) Council Directive 88/642/EEC. OJ L 356, 24.12.1988, p. 74

European Union (1991) Commission Directive 91/322/EEC. OJ L 177, 5.7.1991, p. 22

European Union (1995) Commission Directive 95/320/EC. OJ L 188, 9.8.1995

Henschler D (1985) Development of occupational limits in Europe. Ann Am Conf Ind Hyg 12: 37–40

Lehmann KB (1886) Experimentelle Studien über den Einfluß technisch und hygienisch wichtiger Gase und Dämpfe auf den Organismus. Arch Hyg 5, 1–226

Lehnert G, Schaller KH (Eds) (2000) Biologisches Monitoring in der Arbeitsmedizin. Kompendium von Beiträgen der Mitglieder der Arbeitsgruppe Aufstellung von Genzwerten im biologischen Material der Senatskommission zur Prüfung gesundheitsschädlicher Arbeitsstoffe der Deutschen Forschungsgemeinschaft. Gentner Verlag, Stuttgart

SCOEL (1998) Occupational exposure limits. Recommendations of the Scientific Committee for Occupational Exposure Limits to Chemical Agents 1994–1997. Luxembourg: Office for Official Publications of the European Communities

SCOEL (1999) Scientific Committee Group on Occupational Exposure Limits: Methodology for the derivation of occupational exposure limits. Key documentation. EUR 19253; European Commission, DG V/F.5, Luxembourg: Office for Official Publications of the European Communities [ISBN 92-828-8106-7]

SCOEL (2003) Occupational exposure limits. Recommendations of the Scientific Committee for Occupational Exposure Limits to Chemical Agents. Luxembourg: Office for Official Publications of the European Communities

Thier R, Bolt HM (2001) European aspects of standard setting in occupational hygiene and medicine. Rev Environ Health 16: 81–86

Authors: H.M. Bolt, R. Thier

BAT Value Documentation

Cyclohexane

BAT	170 mg total 1,2-cyclohexanediol/g creatinine Sampling time: end of exposure or end of shift, with long-term exposure after several previous shifts
Date of evaluation	2001
CAS No.	110-82-7
Synonyms	Hexahydrobenzene Hexamethylene Hexanaphthene
Chemical name (CAS)	Cyclohexane
Formula	C_6H_{12}

Molecular weight	84.16
Melting point	6.5°C
Boiling point	80.8°C
Density at 25°C	0.78 g/cm^3
Vapour pressure at 20°C	104 hPa
Solubility in water at 20°C	50 mg/l
MAK [last established: 2001]	200 ml/m^3 ≙ 700 mg/m^3

Cyclohexane is a colourless, highly inflammable liquid with a high vapour pressure of 104 hPa at 20°C. It is insoluble in water, but readily soluble in alcohols, hydrocarbons, ethers or chlorinated hydrocarbons.

It is used mainly as an intermediate in the production of nylon. It is used in the production of cyclohexanol and cyclohexanone, and as a solvent for lacquers, resins, glues and for the extraction of ethereal oils. Cyclohexane does not usually occur as the pure product, but is used industrially in accompaniment with other substances. Mainly cyclohexane–toluene mixtures are used. There are therefore only few data available from field studies for pure cyclohexane at the workplace.

1 Toxicokinetics

1.1 Absorption and distribution

Cyclohexane is absorbed mainly by inhalation; a large proportion is exhaled unchanged (Sandmeyer 1981). Pulmonary retention of the substance was given as 18.4 % (resting conditions) and 34 % (workplace conditions) (Mutti et al. 1981). Cyclohexanol in urine, however, represents only about 1 % of the absorbed dose.

Wetting an area of skin of 360 cm^2 (about 2 % of the body surface) with a saturated solution of cyclohexane in water did not lead to any significant increase in the internal exposure to the substance (Fiserova-Bergerova et al. 1990).

In persons exposed to cyclohexane in concentrations of 350 mg/m^3 and up to 2500 mg/m^3, respectively, the cyclohexane concentrations in the ambient air and those in alveolar air correlated linearly (Mutti et al. 1981, Perbellini et al. 1980).

Oil/air and blood/air distribution coefficients in man were given as 293 and 1.3 (Perbellini et al. 1985). The corresponding tissue/air coefficients were found to be 11 in the liver, 7 in the kidneys, 10 in the brain, 11 in muscle, 6 in the heart, 3 in the lung and 260 in body fat. These data indicate that after cyclohexane has been incorporated it is distributed mainly in adipose tissue.

1.2 Metabolism

The metabolism of cyclohexane to cyclohexanone has been described (Elliott et al. 1959, Mráz et al. 1994, Perbellini and Brugnone 1980, Yasugi et al. 1994) and is shown in Figure 1. Data for the biotransformation of cyclohexane in man are to date inadequate. After volunteers were exposed to cyclohexane concentrations of 1050 mg/m^3 for 8 hours, over a period of 72 hours 0.5 % was excreted in the urine as cyclohexanol, 23.4 % as 1,2-cyclohexanediol and 11.3 % as 1,4-cyclohexanediol, determined as the sum of the free and conjugated compounds.

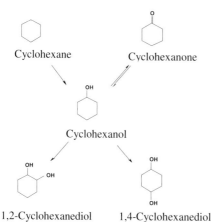

Figure 1. The metabolism of cyclohexane in man

1.3 Elimination

According to data from the literature, alveolar exhalation of the substance nine hours after exposure does not exceed 10 % of the total amount absorbed (Mutti et al. 1981). In animal experiments cyclohexane is excreted with the urine partly in unchanged form and after biotransformation in the liver and in the kidneys in the form of its metabolites. Renal excretion of the three cyclohexane metabolites, cyclohexanol, 1,2-cyclohexanediol and 1,4-cyclohexanediol, was found to be 0.5 %, 23.4 % and 11.3 %, respectively, of the total dose of cyclohexane absorbed (Mráz et al. 1994). Cyclohexanol and the two diols are excreted in man both in glucuronidized and also in free, non-conjugated form (Trabs 1999). As a result of further oxidative biotransformation the amount of cyclohexanol renally excreted is small; only about 10 % of the amount of 1,2-cyclohexanediol excreted (Mráz et al. 1998, Trabs 1999, Walter et al. 1999). The resulting elimination half-times for 1,2-cyclohexanediol were found to be on average 17.0 ± 5.2 hours (Mráz et al. 1998) and 15.8 ± 1.9 hours (Trabs 1999, Walter et al. 1999). The half-time for 1,4-cyclohexanediol determined by Mráz et al. 1998 is of a comparable order of magnitude, while Trabs determined a half-time that was around twice as high, with an average value of 31.5 hours (Trabs 1999). This indicates that both metabolites accumulate. Cyclohexanol, however, was found to be eliminated with mean half-times of 4.6 hours (Trabs 1999, Walter et al. 1999) and 1.5 hours (Mráz et al. 1998). For the monitoring of occupationally exposed persons, the cyclohexane metabolites 1,2-cyclohexanediol and 1,4-cyclohexanediol should therefore be determined at the end of the shift after several previous shifts.

2 Critical Toxicity

2.1 Acute toxicity

Cyclohexanol is of low acute toxicity (Sandmeyer 1981). It has depressive effects on the central nervous system and is narcotic in high concentrations.

2.2 Chronic toxicity

High vapour concentrations are reported to cause irritation of the mucous membranes of the eyes and the respiratory tract, which produce drowsiness, dizziness, nausea, coughing and vomiting, and at extremely high concentrations can provoke respiratory arrest, unconsciousness and collapse. Further details are at present not available. The cyclohexanol metabolite cyclohexanone was classified as carcinogenic in Category 3 (now Carcinogen category 3B) in 1994 and the MAK value of 50 ml/m^3 was withdrawn (Greim 1994).

In a field study, in which 33 female workers from the glue industry were subdivided into a group of 17 women with low exposure (5 to 13 mg/m^3) and a group of 16 women with high exposure (15 to 274 mg/m^3), in neither subgroup were symptoms described. Laboratory analysis did not yield any unusual findings (Yasugi et al. 1994).

3 Exposure and Effects

Field studies and the results of laboratory analyses are available for the external and internal exposure to cyclohexane (e.g. Mráz et al. 1994, 1998, Mutti et al. 1981, Ong et al. 1991, Perbellini et al. 1980, Perbellini and Brugnone 1980, Perico et al. 1999, Trabs 1999, Walter et al. 1999, Yasugi et al. 1994). In particular the field study of Perico et al. (1999), a simulation model of the same author based on data of Perbellini et al. (1988, 1990) and Mráz et al. (1994), and laboratory studies of Mráz et al. (1998), Trabs (1999) and Walter et al. (1999) can be used as the basis for the evaluation of a BAT value for cyclohexane. Table 1 compares the results of the studies. The cyclohexane metabolites cyclohexanol and in particular 1,2-cyclohexanediol and 1,4-cyclohexanediol are suggested in all of the studies as possible exposure parameters after exposure to cyclohexane. Similar metabolite ratios were found in all.

4 Selection of the Indicators

In principle, the solvent itself and all the metabolites formed that are detectable in urine, but in particular the parameters named below, are suitable for the biological monitoring of workers exposed to cyclohexane:
– the cyclohexane concentration in alveolar air
– the cyclohexane concentration in blood
– the cyclohexanol concentration in urine
– the 1,2-cyclohexanediol and 1,4-cyclohexanediol concentration in urine

Cyclohexanol and the two isomeric diols are excreted in free and in glucuronidized form. As the critical toxicity is based on cyclohexanol, this metabolite is of particular importance. Parameters for monitoring the effects are at present unknown.

Table 1. Field studies, PBPK modelling and experiments in man to determine the correlation between external exposure to cyclohexane and the internal exposure on the basis of the cyclohexane metabolites excreted

References	Study, exposure conditions	Air cyclohexane (mg/m³)	Urine cyclohexane metabolites (mg/g creatinine)			Comments
			1,2-cyclo-hexanediol	1,4-cyclo-hexanediol	cyclo-hexanol	
Perico et al. 1999	**field study in a shoe factory** (N = 156 workers) 5 workdays (8 h/day)	102 ± 115 (7–617) 120 ± 128 (Monday) 80 ± 92 (Thursday) 700	13.7 27.8 60.6	8.4 14.1 36.8		exposure to a mixture of solvents y(1,2-diol) = 0.081x (CH$_{Air}$) + 3.9 y(1,4-diol) = 0.049x (CH$_{Air}$) + 2.5 *extrapolated to the MAK value of 700 mg/m³*
Perico et al. 1999 (with data from Perbellini et al. 1988, 1990 and Mráz et al. 1994)	**simulation model (PBPK)** 5 days (8 h/day), physical activity: AV = 15 l/min, Q = 6 l/min AV = 10 l/min, Q = 10 l/min AV = 15 l/min, Q = 6 l/min AV = 10 l/min, Q = 10 l/min	1030 700	day 1: 200 150 day 1: 136 102	day 5: 350 250 day 5: 238 170		PBPK calculated for normal and increased physical activity (expressed as AV and Q) *extrapolated to the MAK value of 700 mg/m³*
Mráz et al. 1998	**laboratory study** (N = 8 persons) 8 hours, at rest	1010 700	136.9 95	71.4 50	12.9 9	excretion of the metabolites determined 6–8 hours after the beginning of exposure; urinary excretion increased on days 2 and 3 of exposure by about 40 % and 60 % *extrapolated to the MAK value of 700 mg/m³*
Trabs 1999, Walter et al. 1999	**laboratory study** (N = 17 persons) 8 hours, physical activity of 75 watts, 10 min/h	1015 ± 77 700	191.4 ± 107.4 132.0 (170)	88.5 ± 56.5 61.0 (85)	16.5 ± 9.2 11.4 (20)	excretion of the metabolites determined 8–10 hours after the beginning of exposure *extrapolated to the MAK value of 700 mg/m³ (95th percentile)*

AV = alveolar ventilation; Q = flow rate of the blood; 1,2-diol = 1,2-cyclohexanediol; CH = cyclohexane; 1,4-diol = 1,4-cyclohexanediol

4.1 Cyclohexane in alveolar air

The available data do not allow any statement to be made about the extent to which alveolar air is a diagnostically valid parameter for detecting exposure to cyclohexane. In addition, there is no standardized sampling procedure available for routine practice (Lauwerys and Hoet 1993, Mutti *et al.* 1981, Perbellini *et al.* 1979, 1980, Rosenberg *et al.* 1989).

4.2 Cyclohexane in blood

In the standardized test procedures of the studies of Trabs (1994) and Walter *et al.* (1999), defined sampling times were set to determine the concentrations and kinetics of cyclohexane in blood during the invasion, equilibrium and elimination phases (Trabs 1999, Walter *et al.* 1999). The data collected yielded differentiated information e.g. about differences in the invasion and evasion kinetics, the steady state and the calculation of biological half-times. The invasion of cyclohexane in the blood increases linearly in the first half hour with a correlation coefficient of $r = 0.90$. During the exposure period, the cyclohexane concentration reaches a plateau one hour after the beginning of exposure with a more or less constant concentration during the determination period. The mean cyclohexane concentration during this steady state is 1.23 ± 0.32 mg/l.

On the other hand, in another field study with external exposure to cyclohexane concentrations of 300 ml/m^3 with a fluctuation range of 17 to 2484 mg/m^3, a cyclohexane concentration in blood of about 0.50 mg/l was found (Perbellini and Brugnone 1980). With the support of another field study, the authors recommended a cyclohexane value of 0.44 mg/l blood as the parameter in biological material, which correlates with a cyclohexane concentration of 300 ml/m^3 (Brugnone and Perbellini 1985). The concentrations determined after occupational and experimental exposure to cyclohexane therefore differ by a factor of about 2.5. Experimentally determined half-times for the elimination of cyclohexane from the blood show that a biphasic course of elimination must be taken into account (Trabs 1999). In the first hour after the end of the standardized exposure, cyclohexane is eliminated from the blood with a half-time of 0.39 ± 0.09 hours. After this period elimination takes place with a half-time of 1.24 ± 0.36 hours. In individual cases the determination of the blood cyclohexane concentration can provide valuable evidence of incorporation, as it reflects, as a result of the short invasion and elimination times, the current external exposure situation. This parameter is less suitable for routine investigations and in particular as the basis for the evaluation of a threshold limit value.

4.3 Cyclohexane metabolites in urine

The metabolites cyclohexanol, 1,2-cyclohexanediol and 1,4-cyclohexanediol can be determined simultaneously. The main advantages of the determination of the metabolites in urine are that
- cyclohexanol and the isomeric diols are not detectable in the urine of persons not occupationally exposed to cyclohexane
- the concentrations of the cyclohexane metabolites, in particular those of the diols, are comparatively high
- the metabolite concentrations in urine are proportional to the external exposure
- these metabolites can also be used as biological parameters for occupational exposure to cyclohexanone and cyclohexanol.

5 Methods

The gas chromatographic head space method is a suitable and tested procedure for the determination of cyclohexane in alveolar air and blood. It is carried out in analogy to the quantification of cyclohexanone in blood (Angerer and Schaller 1983). Gas chromatographic methods of analysis are particularly suitable for the determination of the cyclohexane metabolites in urine. The working group "Analyses of Hazardous Substances in Biological Materials" of the DFG Commission for the Investigation of Health Hazards of Chemical Compounds in the Work Area has tested and approved a capillary gas chromatographic procedure with flame ionization detector for the determination of cyclohexanol and of 1,2-cyclohexanediol and 1,4-cyclohexanediol in urine.

After acid hydrolysis and the addition of an internal standard, the cyclohexane metabolites are extracted with ethyl acetate. The analytical detection limit is around 1 mg of each component per l urine. Analytical determination of the metabolites includes both the conjugated fraction in the urine and also that in unbound form. The above procedure does not allow the separation of the metabolically formed cis/trans isomers 1,2-cyclohexanediol and 1,4-cyclohexanediol, or only in negligible amounts, so that the determination also takes into account the total amount of both components.

6 Background Exposure

Cyclohexane is a component of some glues and resins. The indoor survey 1985/86 of the German Federal Environmental Agency yielded on the basis of 479 indoor determinations a median value of 6 µg/m^3 and a 95th percentile of 18 µg/m^3 (Krause et al. 1987). The highest value determined was 590 µg/m^3. In urine samples of the general population cyclohexane metabolites were not detected with the usual analytical methods. Using a sensitive method, Perico et al. (1999) determined mean amounts of 1,2-cyclohexanediol excreted in the urine of 31 workers not exposed to the substance of 0.5 mg/g creatinine, with a median of 0.3 mg/g creatinine.

7 Evaluation of the BAT Value

The studies shown in Table 1 were used to evaluate a BAT value for cyclohexane. In these studies the parameters used are the excretion of the metabolites 1,2-cyclohexanediol and 1,4-cyclohexanediol. In addition in Table 1 the amounts of cyclohexanol excreted in urine are given. From the study of Perico et al. (1999) relationships between external and internal exposure can be deduced. A cyclohexane concentration in air of 300 ml/m^3 corresponds to metabolite concentrations in urine of about 61 mg 1,2-cyclohexanediol/g creatinine and 37 mg 1,4-cyclohexanediol/g creatinine (Perico et al. 1999).

The results of biomonitoring demonstrate the cumulative properties of both metabolites. In a simulation model also described by Perico et al. (1999), which includes the laboratory studies of Mráz et al. (1994) and Perbellini et al. (1988, 1990), at the end of days 1 and 5 of exposure, with alveolar ventilation of 15 l/min and after extrapolation to 300 ml/m^3, 1,2-cyclohexanediol concentrations were calculated of 136 mg and 238 mg/g creatinine.

The two experimental studies in man of Mráz et al. (1998), and Trabs (1999) and Walter et al. (1999), were carried out with comparable external exposure to cyclohexane of about 300 ml/m^3. The results of the two studies are of a similar order of magnitude.

In the study of Trabs (1999) there are data available for the 95th percentile of the metabolite concentrations which can be used in the evaluation of a BAT value. On this basis the BAT value has been set at

170 mg total 1,2-cyclohexanediol/g creatinine

Sampling of the urine should be carried out at the end of the shift after several previous shifts.

8 References

Angerer J, Schaller KH (1983) Head-Space-Technik, Sammelmethode. in: Henschler D (Ed.) Analytische Methoden zur Prüfung gesundheitsschädlicher Arbeitsstoffe, Band 2: Analysen in biologischem Material, 5. Lieferung, VCH, Weinheim

Brugnone F, Perbellini L (1985) Biological monitoring of occupational exposure to solvents by analysis of alveolar air and blood. in: WHO Regional office for Europe, Nordic Council of ministers (Ed.) Chronic effects of organic solvents on the central nervous system and diagnostic criteria, Report on a joint WHO/Nordic Council of Ministers working group, Copenhagen, 56–89

Elliott TH, Parke DV, Williams RT (1959) Studies in detoxication: the metabolism of cyclo[^{14}C]hexane and its derivatives. Biochem J 72: 193–200

Fiserova-Bergerova V, Pierce JT, Droz PO (1990) Dermal absorption potential of industrial chemicals: criteria for skin notation. Am J Ind Med 17: 617–635

Greim H (Ed.) (1998) Cyclohexanone. in: Occupational toxicants, Vol. 10, Wiley-VCH, Weinheim

Krause C, Mailahn W, Nagel R, Schulz C, Seifert B, Ullrich D (1987) Occurrence of volatile organic compounds in the air of 500 homes in the Federal Republic of Germany. in: Indoor air '87: Proceedings of the 4th international conference on indoor air quality and climate, Volume 1, Institute for Soil, Water and Air Hygiene, Berlin, 102–106

Lauwerys RR, Hoet P (1993) Industrial chemical exposure: guidelines for biological monitoring. Biological monitoring of exposure to organic substances. Lewis Publishers, Boca Raton–Ann Arbor–London–Tokyo, 105–108

Mráz J, Gálová E, Nohová H, Vitková D (1994) Markers of exposure to cyclohexanone, cyclohexane and cyclohexanol: 1,2- and 1,4-cyclohexanediol. Clin Chem 40: 1466–1468

Mráz J, Gálová E, Nohová H, Vitková D (1998) 1,2- and 1,4-cyclohexanediol: major urinary metabolites and biomarkers of exposure to cyclohexane, cyclohexanone and cyclohexanol in humans. Int Arch Occup Environ Health 71: 560–565

Mutti A, Falzoi M, Lucertini S, Cavatorta A, Franchini I, Pedroni C (1981) Absorption and alveolar excretion of cyclohexane in workers in a shoe factory. J Appl Toxicol 1: 220–223

Ong CN, Sia GL, Chia SE, Phoon WH, Tan KT (1991) Determination of cyclohexanol in urine and its use in environmental monitoring of cyclohexanone exposure. J Anal Toxicol 15: 13–16

Perbellini L, Brugnone F, Fracasso ME, Leone R (1979) Studio sulla biotransformazione del cicloesano in ratti e nell' uomo. Il Convegno sulla Patologia indotta dai tossici ambientali ed occupazionali Turin: 21–22

Perbellini L, Brugnone F (1980) Lung uptake and metabolism of cyclohexane in shoe factory workers. Int Arch Occup Environ Health 45: 261–269

Perbellini L, Brugnone F, Pavan I (1980) Identification of the metabolites of n-hexane, cyclohexane, and their isomers in men's urine. Toxicol Appl Pharmacol 53: 220–229

Perbellini L, Brugnone F, Caretta D, Maranelli G (1985) Partition coefficients of some industrial aliphatic hydrocarbons (C5–C7) in blood and human tissues. Br J Ind Med 42: 162–167

Perbellini L, Mozzo P, Turri PV, Zedde A, Brugnone F (1988) Biological exposure index of styrene suggested by a physiologico-mathematical model. Int Arch Occup Environ Health 60: 187–193

Perbellini L, Mozzo P, Olivato D, Brugnone F (1990) "Dynamic" biological exposure indexes for n-hexane and 2,5-hexanedione suggested by a physiologically based pharmacokinetic model. Am Ind Hyg Assoc J 51: 356–362

Perico A, Cassinelli C, Brugnone F, Bavazzano P, Perbellini L (1999) Biological monitoring of occupational exposure to cyclohexane by urinary 1,2- and 1,4-cyclohexanediol determination. Int Arch Occup Environ Health 72: 115–120

Rosenberg J, Fiserova-Bergerova V, Lowry LK (1989) Biological Monitoring IV: measurements in urine. Appl Ind Hyg 4: 16–21

Sandmeyer EE (1981) Alicyclic hydrocarbons. in: Clayton GD, Clayton EE (Eds) Patty's industrial hygiene and toxicology, 3rd ed., Wiley-Interscience, New York, 3221–3250

Trabs A (1999) Biologisches Monitoring nach standardisierter Cyclohexan-Einwirkung. Inaugural-Dissertation des Fachbereichs Humanmedizin der Justus-Liebig-Universität Gießen

Walter D, Trabs A, Knecht U, Woitowitz HJ (1999) Toxikokinetische Daten zur Evaluierung eines BAT-Wertes für Cyclohexan. in: Rettenmeier A, Feldhaus Ch (Eds) Dokumentationsband über die 39. Jahrestagung der Deutschen Gesellschaft für Arbeitsmedizin und Umweltmedizin e.V., Rindt-Druck, Fulda, 265–268

Yasugi T, Kawai T, Mizunuma K, Kishi R, Harabuchi I, Yuasa J, Eguchi T, Sugimoto R, Seiji K, Ikeda M (1994) Exposure monitoring and health effect studies of workers occupationally exposed to cyclohexane vapor. Int Arch Occup Environ Health 65: 343–350

Authors: U. Knecht, K.H. Schaller
Approved by the Working Group: 20.02.2001

Formic acid methyl ester

BAT	not yet established
Date of evaluation	2001
Synonyms	Methyl formate Methyl methanoate
CAS No.	107-31-3
Formula	HCO–OCH$_3$
Molecular weight	60.05
Melting point	–98.8°C
Boiling point	31.8°C
Vapour pressure at 20°C	640 hPa
Density at 20°C	0.974 g/ml
MAK [last established: 1996]	50 ml/m^3 ≙ 120 mg/m^3

Formic acid methyl ester is used as a solvent for fatty oils, fatty acids, cellulose esters and acrylic resins. It is also formed as an intermediate and is used as a pesticide against corn weevils.

1 Metabolism and Kinetics

Formic acid methyl ester is absorbed by inhalation, dermally and via the gastrointestinal tract, and after incorporation is hydrolysed to formic acid and methanol. Methanol is also metabolized to formic acid. Thus, from one mol formic acid methyl ester 2 mols formic acid are formed. The half-life of formic acid methyl ester in the body or biological material is unknown. With a value of 2.0 to 2.25 hours, the half-life of methanol in blood after inhalation exposure to methanol is greater than that of formic acid (Sedivec et al. 1981), which was given as 45 to 46 minutes (0.75 hours) by Malorny after oral administration of formic acid and as 45 minutes by Rietbrock after intravenous injection of formate (Malorny 1969, Rietbrock 1969).

In the urine of persons exposed to methanol, a somewhat shorter monophasic half-life than that in blood was given for methanol, namely 1.5 hours (Dutkiewicz et al. 1980). The discrepancy is probably explained by analytical and methodological differences between the two studies.

The metabolism and kinetics of methanol are described in detail in the BAT documentation in Volume 1 of this series (Henschler and Lehnert 1994) and in the MAK documentation (Greim 2001).

2 Critical Toxicity

Formic acid methyl ester is hydrolysed to formic acid and methanol. An important critical metabolite is formic acid, which is regarded as the primary toxic agent (Hayreh et al. 1980, Liesivuori and Savolainen 1991, Martin-Amat et al. 1978, McMartin et al. 1979, Medinsky and Dorman 1995). The neurobehavioural effects described by Sethre et al. seem, however, to be the result of formic acid methyl ester itself or the metabolite methanol (Sethre et al. 1998a).

3 Exposure and Effects

3.1 Effects

Like after exposure to methanol, the systemic toxicity of formic acid methyl ester is caused by the metabolite formic acid. In the usual animal experiments, after treatment with methanol the laboratory animals do not develop the typical toxic symptoms known in man (metabolic acidosis, neuropathy of the optical nerve, visual impairment). In the case of folic acid deficiency, like man, rodents treated with methanol, develop acidosis, which leads to toxic damage to the visual organ (Clay et al. 1975, Dorman et al. 1994, Eells 1991, Lee et al. 1994, McMartin et al. 1975). It must be remembered that the slow phase of metabolism, from formic acid to carbon dioxide and water, is dependent on folic acid (ATSDR 1993). Thus, with an experimentally-induced folic acid deficiency in rodents, a metabolic situation comparable to that of primates and man can be produced, which leads to the accumulation of formic acid and to impairment of the visual system (Lee et al. 1994). The toxicity of methanol in man has been reviewed in detail (Greim 2001).

There are far fewer reports available of the toxic effects of formic acid methyl ester. A few early observations after intoxication or experimental exposure to formic acid methyl ester are described in the MAK documentation. The symptoms observed after accidental occupational exposure to high concentrations of formic acid methyl ester and formic acid ethyl ester, and additionally to methyl acetate and ethyl acetate, correspond to a certain extent with the clinical symptoms after intoxication with methanol (visual disturbances, in some cases even with transient blindness) (Greim 2003). The external application for 20 minutes of an ointment containing formic acid methyl ester caused the death of a 19-month-old child. After alkaline hydrolysis, methanol was detected in the brain and liver (Greim 2003).

There are recent reports of the effects in man after experimental and occupational exposure to formic acid methyl ester (Berode *et al.* 2000, Sethre *et al.* 1998a, 1998b, Sethre *et al.* 2000a, 2000b). In one study, 20 students were exposed in an inhalation chamber to formic acid methyl ester concentrations of 100 ml/m^3 for 8 hours and the results were compared with those for 20 students not exposed. The study was blind. Two of the 20 parameters differed (subjective tiredness, electromyography in a neck muscle) ($p < 0.05$, without alpha-correction). In the multi-variance analysis the influence of time on tiredness was, however, considerably more important than that of exposure. The EMG (electromyography) finding was detected only in one of two neck muscles investigated.

These toxic effects on behaviour observed in persons exposed in inhalation chambers were not observed by the same working group in field studies with occupational exposure to a mixture of substances including low concentrations of formic acid methyl ester (concentration in the air: 36 ± 21, max. 150 ml/m^3) and isopropanol (concentration in the air: 44 ± 16, max. 375 ml/m^3) (Sethre *et al.* 2000b).

3.2 Exposure and effects

In the literature there are only few publications—which, in addition, are from the same working group (Berode *et al.* 2000, Sethre *et al.* 1998a, Sethre *et al.* 2000a, 2000b)—which describe the relationship between external and internal exposure to formic acid methyl ester. The assumed very rapid hydrolysis of formic acid methyl ester could not be confirmed. In volunteers (N = 20) exposed to formic acid methyl ester concentrations of 100 ml/m^3 for 8 hours (range 90–110; Berode *et al.* 2000), the methanol concentration in urine increased from 2.2 ± 0.8 mg/l (before exposure) to 3.5 ± 1.0 mg/l (after exposure) (Sethre *et al.* 2000a). In volunteers (N = 20) exposed to formic acid methyl ester concentrations of 100 ml/m^3 for 8 hours, methanol concentrations of 3.6 ± 1.0 mg/l urine were determined at the end of exposure. The concentrations of methanol in urine before and after exposure were in the range of < 1–2.6 and 2.1–6.4 mg/l, respectively (Berode *et al.* 2000). The exposure-related, slight, but significant increase in the methanol concentration was only 2.6 mg/l urine (95th percentile). On the basis of this study and the known background exposure (the 95th percentile for the background exposure to methanol is given as 3 mg/l urine; see Table 1), the 95th percentile for the methanol excreted in urine can be calculated for exposure to formic acid methyl ester concentrations of 50 ml/m^3, namely 4.3 (3.0 + 1.3) mg/l.

Using a toxicokinetic model, a similar amount was calculated for the methanol excreted in urine (4.5 mg/l) after exposure to formic acid methyl ester concentrations at the workplace of 50 ml/m^3 (Nihlen and Droz 2000).

After exposure to formic acid methyl ester concentrations of 50 ml/m^3, methanol concentrations in urine of 4.3 mg/l are expected; therefore a maximum permissible threshold concentration of 7 mg/l urine has been suggested (Sethre *et al.* 2000). One study described the urinary excretion of methanol and formic acid in 28 workers from two smelting works (N = 9 and N = 19), who were exposed to formic acid methyl ester concentrations of 57.2 ± 45.7 ml/m³ (range 2–156) (Berode et al. 2000). The concentrations of methanol in the urine collected after the shift were < 1–9.7 mg/l (N = 9) and 2.1–15.4 mg/l (N = 19). The authors draw attention, however, to the significant background exposure (< 1.0 – 2.9 (N = 9) and 1.0–33.4 (N = 19) mg methanol/l urine) and suggest formic acid as the metabolite to be used for biological monitoring (Berode et al. 2000). This unusually high background exposure to methanol is, however, the result of the background levels of methanol in urine determined in two workers of 14.2 and 33.4 mg/l, which were probably caused by diet (e.g. formic acid as a preservative or the formation of formic acid from aspartame) (Berode et al. 2000). In persons not occupationally exposed, the 95th percentile for the background exposure to methanol in urine is 3 mg/l; for this reason background levels below this value are to be expected (see Table 1).

In another study, in workers exposed to formic acid methyl ester concentrations of 48 ± 16 ml/m³ (range 0–531 ml/m³) together with isopropanol (22 ± 15 ml/m³), the amount of methanol excreted was determined at the end of the shift (Sethre et al. 1998b). A correlation was found between the methanol concentration in urine and the exposure to formic acid methyl ester (R = 0.9). On the basis of this correlation (see Figure 1), for exposure at the workplace to formic acid methyl ester concentrations of 50 ml/m³, an upper threshold concentration of 4.4 mg methanol/l urine can be calculated (95th percentile). This value correlates well with the amount calculated for the methanol excreted in the studies with volunteers.

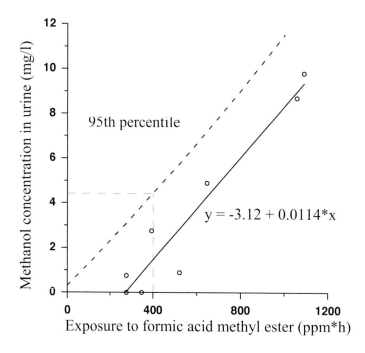

Figure 1. Methanol concentration in urine relative to the exposure to formic acid methyl ester (Sethre et al. 1998b). The straight line was obtained by linear correlation. The dotted line represents the upper threshold concentration for methanol in urine (95th percentile).

4 Background Exposure to Methanol

In persons not occupationally exposed, the concentration of methanol in blood, which is derived both from the diet and from endogenous metabolism, is < 1–2 mg/l (Greim 2003). In alcoholics, as a result of the competitive inhibition of methanol oxidation by ethanol, the methanol levels in blood can be much higher, at around 11 to 27 mg/l.

In persons not occupationally exposed, the 95th percentile for the background exposure to methanol in urine is 3 mg/l (see Table 1).

Table 1. Background exposure to methanol in the urine of persons not occupationally exposed

Methanol in urine (mg/l)		References
mean ± SD (range)	95th percentile	
0.73 (0.32–2.61)	no data	Sedivec et al. 1981
1.1 ± 0.9 (0.6–2.9)	2.95	Heinrich and Angerer 1982
1.90 ± 0.76	3.38	Kawai et al. 1991
1.34	3.86	Ogata and Iwamoto 1990*
1.3 ± 0.8	3.06	Batterman et al. 1998
1.7 ± 0.5	2.75	Sethre et al. 2000a
1.7	no data	Nihlen and Droz 2000

SD: standard deviation
* This study could not be included as the values were below the given detection limit of 3 mg/l.

5 Evaluation of the Indicators and the BAT Value

The metabolite *formic acid* is formed endogenously and is also contained in foodstuffs or is a metabolite of food additives. In 94 randomly selected persons without occupational exposure, formic acid concentrations up to 190 mg/l urine were detected (95th percentile 60 mg/l). The large standard deviation (21 ± 30 mg/l) reflects the great interindividual variability (Heinzow and Ellrott 1992). After occupational exposure to formaldehyde, intraindividual differences in the excretion of formic acid were described that had no explainable cause and no relationship to the external exposure (Schmid et al. 1994). The formic acid concentration in urine is therefore not specific or sensitive enough to be able to quantify the external exposure to formaldehyde, methanol or formic acid methyl ester.

Also *methanol* in the urine of the employees was found to correlate only weakly with the level of formic acid methyl ester in the air. In addition, the methanol concentrations after exposure to formic acid methyl ester at the level of the MAK value are within the range of the background exposure to methanol.

The studies of the working group Sethre–Berode do not allow a BAT value to be evaluated. Many of the questions that arose in the studies were not satisfactorily answered. In particular, no reproducible correlation can be found between formic acid and methanol on the one hand, and methanol or formic acid with formic acid methyl ester in air on the other hand.

The parameters formic acid methyl ester in urine, formic acid methyl ester in blood or formic acid methyl ester in alveolar air may allow better evaluation of the exposure. There are, however, no usable data available.

The data available do not allow a BAT value to be evaluated for the above-mentioned reasons. Also the evaluation of a BAT value for formic acid methyl ester on the basis of its metabolites methanol and formic acid seems questionable, as according to the data available it must now be assumed that the hydrolysis of formic acid methyl ester to methanol and formic acid does not take place very rapidly.

Should a threshold value in biological material be set, this should be (taking into consideration the above-mentioned limitations) between 5 and 10 mg methanol/l urine after several work shifts. 10 mg methanol would be obtained under "worst case" conditions with inhalation exposure alone. Whether percutaneous absorption would lead to a steeper increase in the amount of methanol excreted cannot be deduced with any certainty from the data available.

6 References

ATSDR (Agency for Toxic Substances and Disease Registry) (1993) Methanol toxicity. Am Fam Physician 47: 163–171

Batterman SA, Franzblau A, D'Arcy JB, Sargent NE, Gross KB, Schreck RM (1998) Breath, urine, and blood measurements as biological exposure indices of short-term inhalation exposure to methanol. Int Arch Occup Environ Health 71: 325–335

Berode M, Sethre T, Läubli T, Savolainen H (2000) Urinary methanol and formic acid as indicators of occupational exposure to methyl formate. Int Arch Occup Environ Health 73: 410–414

Clay KL, Murphy RC, Watkins WD (1975) Experimental methanol toxicity in the primate: analysis of metabolic acidosis. Toxicol Appl Pharmacol 34: 49–61

Dorman DC, Moss OR, Farris GM, Janszen D, Bond JA, Medinsky M (1994) Pharmacokinetics of inhaled [^{14}C] methanol and methanol-derived [^{14}C] formate in normal and folate-deficient cynomolgus monkeys. Toxicol Appl Pharmacol 128: 229–238

Dutkiewicz B, Konczalik J, Karwacki W (1980) Skin absorption and per os administration of methanol in men. Int Arch Occup Environ Health 47: 81–88

Eells JT (1991) Methanol-induced visual toxicity in the rat. J Pharmacol Exp Ther 257: 56–63

Greim H (Ed.) (2001) Methanol. in: Occupational toxicants, Vol. 16, Wiley-VCH, Weinheim

Greim H (Ed.) (2003) Methylformiat. Gesundheitsschädliche Arbeitsstoffe, Toxikologisch-arbeitsmedizinische Begründungen von MAK-Werten, 36. Lieferung, Wiley-VCH, Weinheim

Hayreh MMS, Hayreh SS, Baumbach GL, Cancilla P, Martin-Amat G, Tephly TR (1980) Ocular toxicity of methanol: an experimental study. in: Merigen WH, Weiss B (Eds) Neurotoxicity of the visual system, Raven Press, New York, 35–53

Heinrich R, Angerer J (1982) Occupational chronic exposure to organic solvents. X. Biological monitoring parameters for methanol exposure. Int Arch Occup Environ Health 50: 341–349

Heinzow B, Ellrott T (1992) Ameisensäure im Urin – ein sinnvoller Parameter der umweltmedizinischen Diagnostik? Zentralbl Hyg Umweltmed 192: 455–461

Henschler D, Lehnert G (Eds) (1994) Methanol. in: Biological exposure values for occupational toxicants and carcinogens – Critical data evaluation for BAT and EKA values, Vol. 1, VCH, Weinheim, pp. 99–114

Kawai T, Yasugi T, Mizunuma K, Horiguchi S, Hirase Y, Uchida Y, Ikeda M (1991) Methanol in urine as a biological indicator of occupational exposure to methanol vapor. Int Arch Occup Environ Health 63: 311–318

Lee EW, Garner CD, Terzo TS (1994) Animal model for the study of methanol toxicity: comparison of folate-reduced responses with published monkey data. J Toxicol Environ Health 41: 71–82

Liesivuori J, Savolainen H (1991) Methanol and formic acid toxicity: biochemical mechanisms. Pharmacol Toxicol 69: 157–163

Malorny G (1969) Die akute und chronische Toxizität der Ameisensäure und ihrer Formiate. Z Ernährungswiss 9: 332–339

Martin-Amat G, McMartin KE, Hayreh SS, Hayreh MS, Tephly TR (1978) Methanol poisoning: ocular toxicity produced by formate. Toxicol Appl Pharmacol 45: 201–208

McMartin KE, Makar AB, Martin A, Palese M, Tephly TR (1975) Methanol poisoning. I. The role of formic acid in the development of metabolic acidosis in the monkey and the reversal by 4-methylpyrazole. Biochem Med 13: 319–333

McMartin KE, Martin-Amat G, Noker PE, Tephly TR (1979) Lack of a role for formaldehyde in methanol poisoning in the monkey. Biochem Pharmacol 28: 645–649

Medinsky MA, Dorman DC (1995) Recent developments in methanol toxicity. Toxicol Lett 82/83: 707–711

Nihlen A, Droz PO (2000) Toxicokinetic modelling of methyl formate exposure and implications for biological monitoring. Int Arch Occup Environ Health 73: 479–487

Ogata M, Iwamoto T (1990) Enzymatic assay of formic acid and gas chromatography of methanol for urinary biological monitoring of exposure to methanol. Int Arch Occup Environ Health 62: 227–232

Rietbrock N (1969) Kinetik und Wege des Methanolumsatzes. Naunyn-Schmiedebergs Arch Exp Pathol Pharmakol 263: 88–105

Schmid K, Schaller KH, Angerer J, Lehnert G (1994) Untersuchungen zur Dignität der Ameisensäureausscheidung im Urin für umwelt- und arbeitsmedizinische Fragestellungen. Zentralbl Hyg Umweltmed 196: 139–152

Sedivec V, Mraz M, Flek J (1981) Biological monitoring of persons exposed to methanol vapours. Int Arch Occup Environ Health 48: 257–271

Sethre T, Läubli T, Hangartner M, Krüger H (1998a) Neurobehavioral effects of experimental methylformate exposure. Occup Hyg 4: 321–331

Sethre T, Läubli T, Riediker M, Hangartner M, Krüger H (1998b) Neurobehavioral effects of low level solvent exposures in a foundry. Cent Eur J Occup Environ Med 4: 316–327

Sethre T, Läubli T, Berode M, Hangartner M, Krüger H (2000a) Experimental exposure to methyl formate and its neurobehavioral effects. Int Arch Occup Environ Health 73: 401–409

Sethre T, Läubli T, Hangartner M, Berode M, Krüger H (2000b) Isopropanol and methylformate exposure in a foundry: exposure data and neurobehavioural measurements. Int Arch Occup Environ Health 73: 528–536

Authors: H. Drexler, G. Csanády
Approved by the Working Group: 11.01.2002

Hexachlorobenzene, Addendum

BAT	150 µg hexachlorobenzene/l plasma or serum Sampling time: no restrictions
Date of evaluation	2002

1 Re-evaluation of the BAT Value

Hexachlorobenzene is a substance with typical tumour-promoting properties. The liver, bile duct and kidney tumours observed in experimental animals may be attributed to the cytochrome P450-mediated formation of oxygen radicals resulting in cell damage and subsequent compensatory hyperplasia. Thyroid tumours result from the induction of glucuronyl transferases which causes increased glucuronidation and excretion of thyroxine and subsequent enlargement of the thyroid gland. As for the majority of substances with tumour-promoting properties, any clastogenic effects of hexachlorobenzene are very weak. As a result of its tumour-promoting properties, hexachlorobenzene was classified in Carcinogen category 4 in 1998. Derivation of a MAK value for the substance on the basis of the limited data available for effects in man was not possible (Greim 2001). A scientifically grounded BAT value was, however, available (see the BAT documentation for hexachlorobenzene in Volume 2 of the present series; Greim and Lehnert 1995), and re-evaluation was therefore necessary.

1.1 Metabolism and kinetics

In rhesus monkeys and rats, pentachlorophenol was found to be the main metabolite of hexachlorobenzene in urine. In addition, tetrachlorohydroquinone, pentachlorothiophenol, pentachlorothioanisole, 2,4,5-trichlorophenol, 1,2,4,5-tetrachlorobenzene, 2,3,4,6-tetrachlorobenzene and other unidentified chlorophenols and chlorobenzenes occurred as metabolites (see the BAT documentation for hexachlorobenzene in Volume 2).

In 100 persons of a Spanish population (Flix, Tarragona province) exposed to hexachlorobenzene, concentrations in serum were determined in the range of 1.1 to 953 µg hexachlorobenzene/l. In urine, pentachlorophenol levels between 0.58 and 13 µg per 24 hours were detected. A glutathione conjugate was identified as the main metabolite in urine, and was identified after hydrolysis as pentachlorobenzothiol, with levels between 0.18 and 84.0 µg per 24 hours. The amounts of pentachlorobenzothiol excreted in urine correlated closely with the hexachlorobenzene concentration in serum. An increase in the hexachlorobenzene concentration in serum of 1 µg/l led to the excretion of 2.12 µg pentachlorobenzothiol in urine over 24 hours in men and of 0.67 µg in women. The relationship between the hexachlorobenzene concentration in serum and the pentachlorophenol concentration in urine was weaker and only significant in men (To-Figueras *et al.* 1997).

In a follow-up study with 53 persons, in addition to the urine also the faeces were investigated for hexachlorobenzene and its metabolites. In all persons hexachlorobenzene was detected in the faeces; the concentrations were on average 395.4 ng/g faeces dry weight (range 11–3025 ng/g) and correlated closely with the hexachlorobenzene concentration in serum (124.2 µg/l). In 51 % of the persons pentachlorophenol was detected in the faeces (mean value 12.3 ng/g, range 5–70 ng/g) and in 54 % of the persons pentachlorobenzothiol (mean value 32.2 ng/g, range 5–139 ng/g). Pentachlorophenol was determined in the urine of all persons (mean value 3.8 µg in 24 hours, range 0.6–18.0 µg in 24 hours) as was pentachlorobenzothiol (mean value 8.8 µg in 24 hours, range 0.5–86.9 µg in 24 hours). Other metabolites were not detected. The authors estimated that about 3 % of the total hexachlorobenzene in blood is excreted daily with the urine as metabolites and 6 % with the faeces as unchanged hexachlorobenzene (To-Figueras *et al.* 2000).

Figure 1 shows the metabolism of hexachlorobenzene in man.

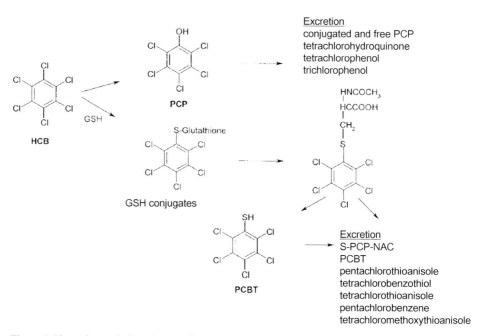

Figure 1. The main metabolic pathways of hexachlorobenzene (HCB) and the main metabolites in urine and faeces (from To-Figueras *et al.* 1997). PCP: pentachlorophenol, GSH: reduced glutathione, PCBT: pentachlorobenzothiol, S-PCP-NAC: *S*-pentachlorophenyl-*N*-acetyl-*L*-cysteine

1.2 Mechanism of action

Hexachlorobenzene was shown to induce porphyria in experiments with the rat, mouse, rabbit and pig (BUA 1994). After intoxication of persons occupationally exposed to hexachlorobenzene, disturbances in porphyrin metabolism were observed, with the increased excretion of uroporphyrins and coproporphyrins in urine (Konietzko and Kern 1989). Disturbances in porphyrin metabolism (*porphyria cutanea tarda*) and other signs of intoxication were observed also in the population in south east Turkey who between 1955 and 1959 used seed treated with hexachlorobenzene for making flour and bread (see the BAT documentation in Volume 2 of this series (Greim and Lehnert 1995) and the MAK documentation (Greim 2001)). The cause of the porphyria is thought to be the induction of porphyrin biosynthesis, with a significant increase in delta-aminolaevulinic acid synthetase and inhibition of uroporphyrinogen decarboxylase, by hexachlorobenzene or tetrachlorohydroquinone, a metabolite of pentachlorophenol. The intermediates of haem biosynthesis accumulate in the liver and can be detected in erythrocytes (protoporphyrins) and the urine (coproporphyrins and uroporphyrins) (Edwards *et al.* 1991). There is also evidence that thyroid hormones play an important role in hexachlorobenzene-induced porphyria (Sopena de Kracoff *et al.* 1994).

Pentachlorophenol was found to be the main metabolite of hexachlorobenzene in animal experiments. In man, in particular glutathione conjugates were detected, and pentachlorophenol only in smaller amounts (see Section 1.1; To-Figueras *et al.* 1997, 2000). This metabolic difference may be due to the fact that in animals porphyrinogenic effects are observed even with exposure to relatively low hexachlorobenzene concentrations, but in man only after intoxication. Also that the typical clinical symptoms of pentachlorophenol intoxication, such as hyperpyrexia, tachypnoea, tachycardia and severe perspiration, are not observed in man after exposure to hexachlorobenzene, indicates that pentachlorophenol is not a decisive metabolite for the toxicity in man. In view of this, the sometimes recommended monitoring and limitation of the amount of pentachlorophenol excreted in urine cannot be supported as a method of preventing adverse effects on the health when handling hexachlorobenzene (Ewers *et al.* 1999).

1.3 Exposure and effects

Increased urinary porphyrin levels were observed in 6 persons from a group of 604 persons with median hexachlorobenzene concentrations in serum of 39.8 µg/l. There was, however, no correlation between the hexachlorobenzene concentration and the increased excretion of porphyrin. Also in the 15 factory workers with the highest hexachlorobenzene concentrations (> 250 µg/l; maximum value 1616 µg/l) there was no correlation with the amount of porphyrin excreted (Herrero *et al.* 1999).

In a Spanish population (Flix; see Section 1.1) who lived in the vicinity of an electrochemical factory, median hexachlorobenzene concentrations in serum of 36.7 µg/l (N = 608) were determined; the 95th percentile was 110 µg/l. The highest serum concentrations were found in male factory workers, with a median value of 54.6 µg/l. Concentrations of other organochlorine compounds were not increased. No significant increase in the risk for adverse effects on the health, chronic diseases, *porphyria cutanea tarda*, thyroid diseases, Parkinson's disease, cancer or impaired reproduction was observed. For factory workers the odds ratio for the prevalence of tumours was not significantly increased (OR 1.9, 95 % confidence interval 0.5, 7.6) (Sala et al. 1999). In an earlier study (Grimalt et al. 1994), the incidences of "tumours of unknown localization", tumours of the thyroid gland, tumours of the brain and soft tissue sarcomas were significantly increased, or borderline, in men who worked in the factory (Greim 2001). 75 male workers employed in this electrochemical factory with a median hexachlorobenzene concentration of 89.3 µg/l ± 190.2 µg/l serum were found to have a concentration of total thyroxin (total T4) of 6.4 µg/dl (2.9–9.8 µg/dl). This was significantly lower than the total T4 level of 17 males never employed in the factory of 7.1 µg/dl (range 5.0–10.0 µg/dl). The median hexachlorobenzene concentration in those 17 males was 14.1 µg/l ± 25.5 µg/l. Also in 11 female workers with a median hexachlorobenzene concentration of 18.8 µg/l ± 8.3 µg/l serum, the concentration of total T4 was significantly decreased (5.5 µg/dl, range 3.8–7.7 µg/dl) compared to that in 89 women who were never employed in the factory (7.4 µg/dl, range 4.2–11.3 µg/dl), with a median hexachlorobenzene concentration of 17.5 µg/l ± 14.2 µg/l serum. In addition, the concentration of γ-glutamyl transferase (γ-GT) was significantly higher in male factory workers (23.0 U/l, range 9.0–255.0 U/l) than in men never employed in the factory (17.0 U/l, range 11.0–44.0 U/l) and, according to the authors, did not change even after adjustment for alcohol consumption. There was also a significant negative relationship between the hexachlorobenzene concentration in serum and the total T4 concentration and a positive relationship with γ-GT (Sala et al. 2001). The clinical relevance of the observed changes is questionable, as the observed changes were in the range of the reference and normal values given by the authors for adults (T4: 4.7–12.4 µg/dl, γ-GT for men 10–50 U/l). In addition, adjustment for possible confounders, such as alcohol consumption, was inadequate. The persons were questioned about alcohol consumption, but the relevant parameters in blood were not investigated. The findings are therefore not suitable for evaluating a BAT value.

There are several publications available about a cohort of Brazilian workers in the production of tetrachloromethane and tetrachloroethene, who were also exposed to hexachlorobenzene as the main component of the residue. The median hexachlorobenzene concentration was 38.8 µg/l serum (range 1–160 µg/l serum). The median age of the workers was 37 years (range 22–54 years) and the median duration of exposure 9 years (range 1–25 years) (Queiroz et al. 1998a).

In a group of 41 exposed Brazilian workers the incidence of micronuclei in the peripheral blood lymphocytes was significantly increased relative to that in a group of 28 workers from other companies (0.9 % compared to 0.25 %). A significant correlation was found between the duration of employment (up to 25 years) and the hexachlorobenzene concentration in serum. There were, however, no significant correlations between the incidence of micronuclei and age, the duration of employment, the hexachlorobenzene concentrations in serum or smoking habits (da Silva Augusto et al. 1997).

In addition, changed immunological parameters were reported in these workers (Queiroz et al. 1997, 1998a, 1998b). However, these effects were not regarded as substance-related as there were no correlations between the parameters investigated and the hexachlorobenzene concentration.

A group of 34 Czech workers who were exposed until 1980 to hexachlorobenzene concentrations of 2.1 to 10.8 mg/m^3 and during the study period between 1983 and 1990 to 0.012 to 0.022 mg/m^3 still had mean serum hexachlorobenzene levels of more than 500 µg/l even 9 years after the exposure concentrations had been reduced (1980: 534 µg/l, 1989: 575 µg/l). In comparison to a control group (N = 100), there were significantly increased values for immunoglobulins (Richter et al. 1994). As a result of the inadequate representation of the results, it is, however, hardly possible to evaluate these findings in the context of the exposure to hexachlorobenzene.

In the BAT documentation from 1984 (see Volume 2) a group of 50 workers from the production of chlorinated hydrocarbons were described, who had mean hexachlorobenzene concentrations of 325 µg/l blood. There were no statistically significant differences relative to the control collective for the biochemical parameters investigated and the liver function values (uroporphyrin and coproporphyrin in urine, lactate dehydrogenase, alkaline phosphatase, total bilirubin, albumin, total protein, ALT, AST, γ-GT, haemoglobin, haematocrit) (Currier et al. 1980).

In a group of 258 workers in Germany employed until 1993 in the production of hexachlorobenzene, the mean hexachlorobenzene concentration in plasma between 1997 and 2001 was 30.3 µg/l and the median concentration 15.0 µg/l. The maximum value was 330 µg/l plasma. In 251 workers (97.3%) with mean hexachlorobenzene concentrations below 150 µg/l plasma, the mean γ-GT concentration was 32.4 U/l, which was in the same range (30.9 U/l) as that for workers with hexachlorobenzene concentrations of below 32 µg/l. Increased γ-GT values were found in workers with higher hexachlorobenzene concentrations. In 3 workers with hexachlorobenzene concentrations between 150 and 200 µg/l plasma and in 4 workers with hexachlorobenzene concentrations of above 200 µg/l plasma, mean γ-GT concentrations of 48.7 U/l and 96.0 U/l were determined (Lewalter, personal communication to the Commission). The results of this study support the observation that changes in liver parameters do not yet occur at hexachlorobenzene concentrations of 150 mg/l serum.

1.4 Re-evaluation of the BAT value

In 1984, on the basis of the occupational-medical studies available, the BAT value for hexachlorobenzene was set at 150 µg/l plasma. Because no statistically significant differences in biochemical parameters such as markers of porphyrin metabolism or liver function were found in the studies, it may be assumed that a BAT value of

150 µg hexachlorobenzene/l plasma

also protects against tumour-promoting effects.

The BAT value has therefore been retained although hexachlorobenzene has been classified in Carcinogen category 4.

2 References

BUA (Beratergremium für umweltrelevante Altstoffe der Gesellschaft deutscher Chemiker) (1994) Hexachlorbenzol, Bericht Nr. 119, Hirzel Verlag, Stuttgart

Currier MF, McClimans CD, Barna-Lloyd G (1980) Hexachlorobenzene blood levels and the health status of men employed in the manufacture of chlorinated solvents. J Toxicol Environ Health 6: 367–377

Edwards IR, Ferry DG, Temple WA (1991) Fungizides and related compounds. in: Hayes WJ Jr, Laws ER Jr (Eds) Handbook of pesticide toxicology, Volume 3, Academic Press, San Diego, 1409–1470

Ewers U, Krause C, Schulz C, Wilhelm M (1999) Reference values and human biological monitoring values for environmental toxins. Int Arch Occup Environ Health 72: 255–260

Greim H (Ed) (2001) Hexachlorobenzene. in: Occupational toxicants, Vol. 16, Wiley-VCH, Weinheim

Greim H, Lehnert G (Eds) (1995) Hexachlorobenzene. in: Biological exposure values for occupational toxicants and carcinogens – Critical data evaluation for BAT and EKA values, Vol. 2, VCH, Weinheim, pp. 59–65

Grimalt JO, Sunyer J, Moreno V, Amaral OC, Sala M, Rosell A, Anto JM, Albaiges J (1994) Risk excess of soft-tissue sarcoma and thyroid cancer in a community exposed to airborne organochlorinated compound mixtures with a high hexachlorobenzene content. Int J Cancer 56: 200–203

Herrero C, Ozalla D, Sala M, Otero R, Santiago-Silva M, Lecha M, To-Figueras J, Deulofeu R, Mascaro JM, Grimalt J, Sunyer J (1999) Urinary porphyrin excretion in a human population highly exposed to hexachlorobenzene. Arch Dermatol 135: 400–404

Konietzko J, Kern G (1989) Hexachlorbenzol. in: Konietzko J, Dupuis H (Eds) Handbuch der Arbeitsmedizin, Ecomed Verlagsgesellschaft, Landberg, München, Zürich, 1. Ergänzungslieferung, IV-2.29.6.6, 1–6

Queiroz ML, Bincoletto C, Perlingeiro RC, Souza CA, Toledo H (1997) Defective neutrophil function in workers occupationally exposed to hexachlorobenzene. Hum Exp Toxicol 16: 322–326

Queiroz ML, Bincoletto C, Perlingeiro RC, Quadros MR, Souza CA (1998a) Immunglobulin levels in workers exposed to hexachlorobenzene. Hum Exp Toxicol 17: 172–175

Queiroz ML, Quadros MR, Valadares MC, Silveira JP (1998b) Polymorphonuclear phagocytosis and killing in workers occupationally exposed to hexachlorobenzene. Immunopharmacol Immunotoxicol 20: 447–454

Richter J, Lauda K, Reznicek J (1994) Immune response in persons occupationally exposed to hexachlorobenzene. Prac Lek 46: 151–154

Sala M, Sunyer J, Otero R, Santiago-Silva M, Ozalla D, Herrero C, To-Figueras J, Kogevinas M, Anto JM, Camps C, Grimalt J (1999) Health effects of chronic high exposure to hexachlorobenzene in a general population sample. Arch Environ Health 54: 102–109

Sala M, Sunyer J, Herrero C, To-Figueras J, Grimalt J (2001) Association between serum concentrations of hexachlorobenzene and polychlorobiphenyls with thyroid hormone and liver enzymes in a sample of the general population. Occup Environ Med 58: 172–177

da Silva Augusto LG, Lieber SR, Ruiz MA, de Souza CA (1997) Micronucleus monitoring to assess human occupational exposure to organochlorides. Environ Mol Mutagen 29: 46–52

Sopena de Kracoff YE, Ferramola de Sancovich AM, Sancovich HA, Kleiman de Pisarev DL (1994) Effect of thyroidectomy and thyroxine on hexachlorobenzene induced porphyria. J Endocrinol Invest 17: 301–305

To-Figueras J, Sala M, Otero R, Barrot C, Santiago-Silva M, Rodamilans M, Herrero C, Grimalt J, Sunyer J (1997) Metabolism of hexachlorobenzene in humans: association between serum levels and urinary metabolites in a highly exposed population. Environ Health Perspect 105: 78–83

To-Figueras J, Barrot C, Sala M, Otero R, Silva M, Delores Ozalla MD, Herrero C, Corbella J, Grimalt J, Sunyer J (2000) Excretion of hexachlorobenzene and metabolites in feces in a highly exposed human population. Environ Health Perspect 108: 595–598

Authors: J. Lewalter, U. Reuter
Approved by the Working Group: 17.06.2002

Lead and its compounds (except lead arsenate, lead chromate and alkyllead compounds)

BAT for women older than 45 years and for men (for women younger than 45 years see Addendum)	400 µg lead/l blood
Date of evaluation	2000
CAS No. (metallic lead)	7439-92-1
Formula (metallic lead)	Pb
Atomic weight (metallic lead)	207.2
Melting point (metallic lead)	327.4°C
Boiling point (metallic lead)	1740°C
Vapour pressure at 970°C (metallic lead)	1.33 mbar
MAK [last established: 2000]	Carcinogen category 3B

The following occupational-medical, toxicological documentation applies for exposure to metallic lead, lead oxides and lead salts. Not taken into account is exposure to covalent lead compounds (e.g. alkyllead compounds), or lead salts with an anion (e.g. lead arsenate, lead chromate), which are carcinogenic or of suspected carcinogenicity.

1 Pharmacokinetics

There are numerous studies available of the absorption, distribution, retention and excretion of lead in the human organism. Particular reference is made to Chamberlain 1985, Skerfving 1993, US Environmental Protection Agency 1986, and WHO 1995.

1.1 Absorption

Lead and its inorganic compounds are absorbed by the human organism via the lungs and gastrointestinal tract. At the workplace inhalation is the most important route of absorption. In 1995 the WHO reviewed the current data for the absorption of lead (WHO 1995).

1.2 Elimination

The absorbed lead is eliminated mainly (75–80 %) with the urine and in smaller amounts via gastrointestinal excretion (WHO 1995). The rate of elimination is relatively slow. It is very difficult to give a biological half-life because of the constantly decreasing availability of the depot in the bones. The "Task Group on Metal Accumulation" gives a half-life for lead in human bones of about 10 years (TGMA 1973).

1.3 Metabolism

The absorbed lead immediately enters the blood stream. About 95 % of the blood lead is bound to the erythrocyte membrane and is distributed in this form in the organism. In the body the lead is present as a displaceable and an irreversibly bound fraction (Baloh 1974).
- The rapid exchange pool represents the biologically effective body burden. It correlates with the blood lead concentration, i.e. the lead concentration in the blood is in equilibrium with the lead concentration in soft tissue.
- Lead readily passes the placental barrier. Foetal blood therefore has more or less the same lead concentration as the maternal blood (Haas et al. 1972, Roels et al. 1978).
- Lead also passes the blood–brain barrier, but is not thought to accumulate in the brain.
- 90 % of the lead bound in the body is contained in the bones and teeth.

2 Critical Toxicity

The literature on the toxicology of lead and its compounds in experiments with animals and in man was reviewed in the occupational-medical, toxicological documentation of the MAK value (Henschler 1977, Greim 2002). There is also extensive data in Environmental Health Criteria 165, "Inorganic lead" from the International Programme on Chemical Safety (IPCS) of the WHO (1995).

In the occupational-medical, toxicological documentation of the MAK value (Greim 2002) the current data for the toxic effects of lead on reproduction, fertility, genotoxicity and carcinogenicity are presented in detail. The MAK documentation also describes the effects on the kidneys, the blood pressure, the haematopoietic system and the thyroid gland in man.

For the setting of the BAT value, the critical effect of lead is the effect on the central nervous system. The most sensitive parameter for the toxic effects of lead and its compounds, with the exception of the alkyl compounds, is the decrease in performance in behavioural tests. The data from current studies relevant for setting the threshold limit value are described below in Section 3.2. For other effects of lead and its compounds on health, the reader is referred to the occupational-medical, toxicological documentation of the MAK value (Henschler 1977, Greim 2002).

3 Exposure and Effects

3.1 Relationship between external and internal exposure

As far as it is known, there is no clearcut relationship between the external and internal exposure to lead. There are many reasons for this. The main reason is that, unlike for the determination of lead in the air (current exposure), the lead concentration in blood represents the long-term uptake of lead and the total body burden. In addition to the inhalation of lead, the substance can be ingested from contaminated hands and with drinks and food. A BAT value cannot, therefore, be evaluated on this basis.

3.2 Relationships between internal exposure and effects

As already explained in Section 2, the BAT value must be evaluated on the basis of the relationship between the blood lead concentration and the neurotoxic effects of lead compounds. The most sensitive parameter for the toxic effects of lead in men and women is the decrease in performance in behavioural tests.

3.2.1 Neurophysiological effects

The most important results published since 1980 from neurophysiological studies of persons exposed to lead at work are described below.

19 studies published between 1970 and 1984 were evaluated in a review (Ehle 1986). Most of the studies report a slight reduction in nerve conduction velocity (about 7 %) at blood lead concentrations below 600 µg/l. Correlation with clinical symptoms was not found in any study. The author emphasises that in most of the studies no relationship was detected between the nerve conduction velocity and blood lead concentrations below 700 µg/l. The reduced nerve conduction velocities in general are of no clinical relevance and are not considered to be a "sub-clinical" manifestation (Ehle 1986).

In a review published in 1990, Davis and Svendsgaard evaluated 32 studies in which various nerve conduction velocities were determined in persons exposed to lead at the workplace. In general, reduced nerve conduction velocities were found. In a meta-analysis the authors combined the results of the studies and calculated the magnitude of the effects. They concluded that the nerve conduction velocity in persons exposed to lead is reduced and the effect is most pronounced in the median nerve (Davis and Svendsgaard 1990). In their analysis the authors draw attention to two longitudinal studies whose results, however, are inconsistent (Seppäläinen et al. 1983, Spivey et al. 1980). While Seppäläinen et al. (1983) described significant reductions in nerve conduction velocity in 89 employees of a factory producing batteries with blood lead concentrations of about 300 to 500 µg/l and an observation period of 4 years, Spivey et al. (1980) could not find any significant differences between control persons and 69 lead smelters (blood lead concentration 600 to 800 µg/l) over a period of 12 to 18 months.

An evaluation of lead and its inorganic compounds by a WHO committee (1995) included a description of the peripheral neurotoxic effects (WHO 1995). The conclusion of the committee that the nerve conduction velocity changes at blood lead concentrations as low as 300 µg/l is based purely on the study of Seppäläinen et al. (1983). The fact that this finding could not be confirmed in other studies is mentioned by the WHO committee, but is not critically analysed from a methodological point of view.

There are only few more recent studies which investigate the relationships between occupational exposure to lead and peripheral nerve function.

Chia et al. (1996a, 1996b) investigated 72 employees from a factory producing batteries; the mean blood lead concentration of the workers was found to be 369 µg/l (range 73 to 685 µg/l). The authors found significant differences between the nerve conduction velocities of the median nerve and ulnar nerve in the employees and those in control persons, without values lower than normal being found on a group basis (Chia et al. 1996a). Clinical findings are not reported. On the basis of the results of the 3-year longitudinal study the authors concluded that a cumulative blood lead concentration of less than 400 µg/l does not cause any neurophysiological changes (Chia et al. 1996b).

Kovala et al. (1997) determined the nerve conduction velocities of the median nerve and sural nerve in 60 workers from a factory producing batteries. The exposure was evaluated by determining the blood lead concentration (mean concentration 331 µg/l, highest concentration 1014 µg/l) over the whole exposure period and the bone lead concentration (tibia) (mean 26 mg/kg). Weak negative correlations, which were not statistically significant, were found between the exposure parameters and the nerve conduction velocity. A threshold value was not given.

There are also studies in which various electrophysiological parameters (evoked potentials, such as e.g. VEP, BAEP, P 300, electrocardiogram) and body sway (posturography) were determined in persons exposed to lead (Chia et al. 1994, Chia et al. 1996c, Fujimura et al. 1998, Kovala et al. 1997, Murata and Araki 1991, Murata et al. 1993, Teruya et al. 1991, Yokoyama et al. 1997). It was shown that lead can influence not only the peripheral nerve fibres, but also the vegetative nervous system, functions of the spinal cord and the central nervous system. It is not possible to make a definitive statement on the clinical relevance of these effects or on whether they are to be considered adverse. Threshold concentrations can also not be defined.

According to a Japanese working group, copper and zinc seem to have antagonistic effects on the sub-clinical effects of lead on peripheral nervous functions (Araki et al.

1987, Murata et al. 1987). With regard to neurophysiological changes, a recent study found that lead mobilized using EDTA is a more sensitive parameter of lead exposure than is the blood lead concentration (Yokoyama et al. 1998).

The polyneuropathy induced by lead progresses with increasing exposure. The course of the disorder depends mainly on the severity of the initial lesion (Triebig 1984). In earlier stages complete regression can be expected within a few months (Krigman et al. 1980, WHO 1995).

In summary, it is possible to detect effects on functions of the peripheral and autonomous nervous system at blood lead concentrations in the range of 300 to 500 µg/l on a group basis, but effects which can be described as adverse occurred consistently only at blood lead concentrations above 600 to 700 µg/l.

3.2.2 Toxic effects on behaviour

3.2.2.1 Database

There are numerous epidemiological studies of the behavioural effects of exposure to lead. The data from the 1980s and 1990s indicate that effects on behaviour can be detected for occupational exposure in the range above 300 µg/l blood.

The following summary takes into account 30 publications; different publications concerning the same study were taken as one, so that in total 24 studies were included. 22 publications provide information on the current effects of lead exposure on the basis of cross-sectional studies, 5 publications describe long-term effects on the basis of longitudinal studies.

3.2.2.2 Documentation of the studies

The studies included in the evaluation are summarized in Tables 1a–d, 2a–2c and 3 at the end of this chapter.

3.2.2.3 Methodological evaluation of the studies

In an evaluation of 28 publications covering 21 studies of exposure to lead, Balbus-Kornfeld et al. (1995) conclude that no definitive statement can be made about the possible adverse effects of cumulative exposure to lead on the central nervous system in man because the extent of cumulative exposure has not been adequately studied. The effects found in the studies may be merely transitory or reversible and are therefore of limited meaning for the evaluation of health effects. This view is not shared by the present authors, as in the argumentation for the setting of a threshold limit value also current effects are seen as potential early signs of long-term effects.

We agree with the above authors, however, that on the basis of the evaluation criteria for studies of behavioural toxicology (DFG 1997) various studies have methodological shortcomings. The comparability of the groups with regard to age, sex and education is taken into consideration by most of the researchers, but there are critical exceptions. In the studies of Haenninen et al. (1978) and Jeyaratnam et al. (1986) the control groups are older, in Williamson and Teo (1986) and Yokoyama et al. (1998) the control groups are younger. Only some of the studies include these group features in the further analysis (Baker et al. 1983, 1984, 1985, Braun and Daigneault 1991, Campara et al. 1984, Chia et al. 1997, Haenninen et al. 1978, Hogstedt et al. 1983, Lindgren et al. 1996, Maizlish et al. 1995, Mantere et al. 1984, Parkinson et al. 1986, Repko et al. 1978, Spivey et al. 1979, Stollery et al. 1989, Valciukas et al. 1978, Zimmermann-Tansella et al. 1983), so that confounding effects cannot be excluded in the other studies.

In the studies of Braun and Daigneault (1991), Campara et al. (1984) Zimmermann-Tansella et al. (1983), Baker et al. (1983, 1984, 1985) and Grandjean et al. (1978) the difference in the results between the exposed persons and the control group in the vocabulary test suggests a different premorbid intelligence. In Baker et al. (1983, 1984, 1985) and Braun and Daigneault (1991) the small difference in the results of this test suggests it is of less importance. The different premorbid intelligence is clearer in the groups investigated by Repko et al. (1978).

Another methodological problem is the comparability of the activities within the exposed groups or between the exposed persons and controls. This aspect is not taken into consideration in many of the studies, which might be due to the fact that the activities of workers are assumed to be comparable. This aspect is included in the evaluation of Stollery et al. (1989), Maizlish et al. (1995), Haenninen et al. (1978), Baker et al. (1983, 1984, 1985), Valciukas et al. (1978), Williamson and Teo (1986), Lindgren et al. (1996), Österberg (1997) and Haenninen et al. (1998). There may be differences between the groups investigated by Campara et al. (1984), Zimmermann-Tansella et al. (1983), Repko et al. (1978), Braun and Daigneault (1991) and Valciukas et al. (1980). The meaning of the variable "activity" for the determination of performance is demonstrated in the study of Lindgren et al. (1996). It was discovered that the duration of employment acts as a suppressor variable, as the duration of employment was connected with further training and thus also promotion to higher positions.

The consumption of alcohol and drugs, and previous illnesses, also as exclusion criteria, were taken into consideration by most of the authors (exceptions: Repko et al. (1978), Banks and Stollery (1988), Dotzauer (1990), Valciukas et al. (1980), Haenninen et al. (1978), Valciukas et al. (1978), Jeyaratnam et al. (1986)). Stollery et al. (1989), Braun and Daigneault (1991), Haenninen et al. (1998), Arnvig et al. (1980) and Williamson and Teo (1986) took only alcohol consumption into consideration.

As a result of the methodological limitations with regard to premorbid intelligence, doubts about the meaningfulness of the studies of Braun and Daigneault (1991), Campara et al. (1984), Zimmermann-Tansella et al. (1983), Baker et al. (1983, 1984, 1985) and Grandjean et al. (1978) are justified. As the importance of premorbid intelligence with regard to its invariance to toxic effects is, however, not completely clear (Balbus-Kornfeld et al. 1995, Haenninen et al. 1998), these studies should be included in the evaluation "with discretion". The study of Repko et al. (1978) should not be included because of differences in premorbid intelligence.

Also in the studies of Dotzauer (1990), Spivey et al. (1979), Valciukas et al. (1978, 1980), Haenninen et al. (1978), Jeyaratnam et al. (1986) and Williamson and Teo (1986) there are methodological shortcomings with regard to the comparability of the groups. There are differences in alcohol consumption, previous illnesses and the age-matching of the persons in the groups. Therefore also these studies should be included in the evaluation only "with discretion". The shortcomings in the evaluation of Banks and Stollery (1988) with regard to alcohol consumption make this study unsuitable for the evaluation of effects of lead.

3.2.2.4 Summary of behavioural effects after occupational exposure to lead

In the evaluation of the meaningfulness of the findings listed in Tables 1a–d and 2a–c, it must be borne in mind that several areas of performance were taken into consideration by most authors, but well-being and subjective complaints were investigated much less often and there are no data on current changes in well-being. The data for symptoms and well-being come from a smaller database and cannot achieve the same significance for the evaluation as the data on performance.

3.2.2.4.1 Effects on performance

Perceptual discrimination was determined specifically in 5 studies, but is covered by numerous other test procedures e.g. in the attention tests. A meta-analysis of the available results for visual interference revealed no significant effects for the groups investigated. The numerous deficits in performance in attention tests do not, however, exclude the possibility that lead-induced impairments in performance can be detected also in tests for perceptual discrimination.

Learning and memory was investigated in 20 studies. Not affected by exposure to lead is long-term memory, i.e. the recollection of things learnt earlier in life. Effects can be seen, however, in the areas of short-term memory and retention of verbal and senseless material. Meta-analysis[1] revealed a marked effect ($d = -0.39$) on the short-term retention of verbal information for an average current blood lead concentration of 400 µg/l. Deficits in performance occur frequently in the active recollection of figural material and remembering numbers. An effect on the remembering of numbers could not be shown after meta-analysis using the overall scores for the memory span (reproducing digits forwards and backwards). The determination of a total score for memory span is open to criticism, because it involves association of different cognitive processes (Lezak 1995).

Concentration and attention were also investigated in 20 studies. With only a few exceptions, vigilance was found to be impaired by exposure to lead. Also complex searching and association performance, as demanded in the digit-symbol test and trail-making test, are affected. The meta-analysis for the digit-symbol test yielded for a group

---

[1] Meta-analysis can be used to summarise studies with the same methods and approaches. The strength of the effect (effect size) for the relationships investigated is determined ($d = 0.20 /50 /80$ as a slight/moderate/strong effect). The negative values here mean that the results in the control group were more favourable than in the exposed group. The analysis was carried out according to the procedure of Fricke and Treinies (1985). For the calculations and results see Meyer-Baron and Seeber (2000)

of studies (average current blood lead concentration 450 µg/l) an effect size of $d = -0.25$, for a second group of studies (average current blood lead concentration 460 µg/l) an effect size of $d = -1.04$.

Other cognitive processes such as verbal expression, encoding and concluding are dealt with in 17 studies. Concluding and visuo-constructive abilities, which are called for e.g. in the Mosaic test, verbal expression and figural intelligence are repeatedly found to be impaired. Meta-analysis revealed an effect size of $d = -0.30$ for visuo-constructive performance at an average current blood lead concentration of 420 µg/l. For verbal expression, as a result of the heterogeneous results, analysis is only possible for a subgroup of studies (average current blood lead concentration 480 µg/l) and yielded a value of $d = -0.58$. For methodological reasons it is not possible to demonstrate an overall effect in the area of logical thinking.

Motor processes were investigated in 19 studies and found with various test procedures to be impaired. In addition to sensorimotor velocity, also dexterity and coordination are affected. For an average current blood lead concentration of 450 µg/l meta-analysis revealed an effect size of $d = -0.20$ for the finger dexterity of the dominant hand.

3.2.2.4.2 Symptoms and changes in personality

Data on well-being and/or complaints are given in 12 studies. The changes recorded include mood and emotions, and neurological, neurovegetative and gastrointestinal symptoms. One of the main areas studied is the complex annoyance, tension, irritability, animosity and conflicts. In addition, adverse effects on drive and subjective loss of memory and concentration are frequently mentioned. The main motor complaints are unsteadiness and weakness in the extremities. Two authors who investigated this aspect noted an increase in accidents at work.

3.2.3 Discussion of the effects

To summarise, the studies repeatedly revealed impairments in performance as a result of exposure to lead at current blood lead concentrations of < 700 µg/l, e.g.
- in the area of learning and memory for short-term retention of verbal information at an average blood lead concentration of 400 µg/l and an effect size of $d = -0.39$,
- in the area of attention and association performance for the digit-symbol test at average blood lead concentrations of 460 µg/l and 480 µg/l, and effect sizes of $d = -1.04$ and $d = -0.25$,
- in the area of more complex cognitive functions for the Mosaic test and verbal expression at an average blood lead concentration of 420 µg/l and $d = -0.30$, and 450 µg/l and $d = -0.58$,
- in the area of sensorimotor functions for finger dexterity at an average of 450 µg/l and $d = -0.20$.

In addition, mood changes, lack of drive and an increase in the number of cases of impairment of motor functions were found.

A more general evaluation of the effects can take either of two directions: it can be based on age-related changes or on statistical reference values. With regard to *age-related changes*, in the study of Campara *et al.* (1984) impaired performance was found in association tests (digit-symbol test) at a current blood lead concentration of 460–600 µg/l corresponding to an age-related change of about 25 years.

With regard to *statistical reference values*, an effect size of $d = -0.20$ means that an increase in the blood lead concentration in the exposed groups relative to the normal value in the control group causes a decrease in the test parameter corresponding to one fifth of the standard deviation of the test performance. For the results of the digit-symbol test in the described meta-analysis, this means that in the studies included in the meta-analysis at a current average blood lead concentration of 460 µg/l a decrease in performance of $d = -1.04$ was found; this corresponds to around one standard deviation in the test score. Theoretically a person's test performance would fall e.g. statistically from the 92nd percentile in the normal distribution to the 24th percentile. Relative to the normal results in the digit-symbol test for the age group, this is indicative of a decrease in performance corresponding to ageing by about 50 years. For the digit-symbol test at current average blood lead concentrations of 480 µg/l (an effect size of $d = -0.25$) the other group of authors found a change corresponding to a quarter of the standard deviation for the normal population. This effect size indicates that a person would fall e.g. from the 92nd percentile to the 82nd. In terms of age, this decrease in performance corresponds to ageing by at least 10 years. If we compare the above with the data of Campara *et al.* (1984), the lead-induced decrease in performance in the digit-symbol test can be regarded as verified and its extent confirmed for a current blood lead concentration in the range of 400–500 µg/l. The other effects can be interpreted in the same way.

As these are not individual findings but effects determined in meta-analyses, the necessity of resetting the threshold limit value is evident. Irrespective of this, the reversibility of the effects is first to be discussed below.

3.2.4 The reversibility of exposure-related changes

The results of studies of the reversibility of the effects are contradictory. Two studies found the effects to be reversible. In the study of Yokoyama *et al.* (1988) the decrease in performance was found to be reversible after the exposure to lead was reduced (blood lead concentration at the time of the first investigation 400–640 µg/l, after 2 years 260–590 µg/l). This result applied to 19 exposed persons. In the study of Baker *et al.* (1985) the reduced exposure (blood lead concentration at the time of the first investigation 500–800 µg/l, after 1 year 370–670 µg/l) led to a marked improvement in the variable mood/emotion without a corresponding change in the performance variables. The authors attributed this to the fact that in the first investigation the symptoms were exaggerated. Because the changes occurred only in the group with the highest level of exposure, the improvements in mood scores could be seen as the result of a reduction in exposure. These effects were seen in 10–12 persons.

It speaks against the reversibility of the effects that in 7 studies dose–response relationships reveal closer correlation in some cases to the accumulated levels (IBL), average exposures (TWA) and to indicators of long-term exposure, such as zinc protoporphyrin (ZPP) than to the current level of lead exposure (Valciukas et al. 1980, Haenninen et al. 1978, Valciukas et al. 1978, Baker et al. 1985, Österberg 1997, Chia et al. 1997, Spivey et al. 1979). This is also clear from the results of Lindgren et al. (1996), which showed that despite the lack of a relationship with the current blood lead concentrations, a dose–response relationship could be determined between performance parameters and the total exposure over the employment period. Also the study of Haenninen et al. (1998) revealed that in the group with higher level exposure the correlation between performance and the cumulative exposure to lead is better than that with the current exposure level, or that a relationship could only be detected to the cumulative exposure to lead. However, this study identifies a closer correlation of the performance in the group with low-level exposure with the current blood lead concentration. Such differentiation is possibly reflected in the results of Maizlish et al. (1995), in which the correlations with the current blood lead concentration are mostly higher.

To summarise, there is evidence from the available studies of clear and long-term behavioural deficits also at current blood lead concentrations < 70 µg/l.

3.2.5 Discussion

It must first of all be noted that none of the studies consider sex-specific differences in the effects. The studies investigated almost exclusively male persons and where (a few) women took part in the study no differentiation is made. Thus, the question of possible sex-specific differences in the effects is not treated in the following discussion.

The mean current blood lead concentration at which effects were detected in performance or personality variables was found to be 310 to about 500 µg/l in the studies cited. When blood lead concentrations determined over a longer period are taken into consideration, values of 290–530 µg/l are associated with statistically verified effects on performance and personality. The study of Lindgren et al. (1996) reports on the lowest levels of exposure with statistically verified effects, but only on a group basis. With mean current exposures of about 280 µg/l, but long-term 400 µg/l, a dose–response relationship was found between the effects on performance and an index of cumulative long-term exposure. Changes in performance affected attention, association and sensorimotor functions. Evidence of effects on performance below the level of 400 µg/l were also found on a group basis in Chia et al. (1997) (exposed versus control groups: 370 versus 60 µg/l) and in Mantere et al. (1984) (310 versus 110 µg/l). All three of the above studies are designated in Tables 1a–c as those which use more than one current blood lead concentration.

If the results of the meta-analysis of Meyer-Baron and Seeber (2002) are taken as the basis for discussing a threshold limit value, the concentration associated with slight and moderate effect sizes for the given performance parameters is in the range of 400–480 µg/l (see Section 3.2.3). The meta-analysis shows that in analogy to age-related changes even these small effect sizes are responsible for a change in test performance corresponding to a 10-year age difference. These changes should be regarded as relevant for health.

Effects on performance that can generally be reproduced begin at an average concentration of 400 µg/l. An evaluation of 30 publications revealed three studies which indicate effects also in the range from about 300 µg/l. In view of this, a BAT value of 400 µg lead/l blood is at present justified. When other publications become available, this value must be reviewed to determine whether and to what extent there is evidence of reproducible effects, in particular in the range between 300 and 400 µg/l.

4 Selection of the Indicators

For several years there has been a consensus of opinion in occupational medicine that the recognition of increased absorption of lead at the workplace must be based mainly on laboratory analysis of specimens from exposed persons. There is a wide spectrum of both exposure and effect parameters available for this purpose. These can be classified as follows:

Parameters of internal exposure
Determination of the lead concentration in biological materials:
- blood lead concentration
- amount of lead excreted with the urine
- amount of lead in bones
- amount of lead excreted with the urine after provocation with complexing agents.

Parameters of effect
Disturbance of porphyrin synthesis is the main biological parameter used to detect the effects of lead. These disturbances are best recognized in man in effects on haematological parameters.
- in blood: δ-aminolaevulinic acid dehydratase (ALA-D), erythrocyte porphyrins (EPs) and zinc protoporphyrins (ZPPs)
- in urine: δ-aminolaevulinic acid (ALA-U).

At present, the lead concentration in blood is the most reliable and practicable parameter for monitoring persons exposed to lead. It is the most specific parameter for evaluating the internal exposure to lead. Effect parameters can no longer be recommended, either for methodological or practical reasons, or for reasons of inadequate sensitivity in view of the fact that the levels of exposure to lead at the workplace are lower than in the past.

5 Methods

Quantitative determination of the lead concentration in whole blood can be carried out using atomic absorption spectrometry, voltammetry and ICP-MS. Reliable and tested methods for these three analytical techniques are included in the series "Analyses of

Hazardous Substances in Biological Materials" (Angerer and Schaller 1985, 1988, 1999) published by the DFG Commission for the Investigation of Health Hazards of Chemical Compounds in the Work Area. Voltammetric analysis is very sensitive, but also very time-consuming and is therefore not used routinely (Ostapczuk 1992). Because special apparatus is needed, ICP-MS is also not used routinely (Delves and Campbell 1988, Gercken and Barnes 1991).

Today graphite furnace atomic absorption spectrometry (GF-AAS) with Zeeman compensation is used almost exclusively. Usually, after dilution and the addition of a matrix modifier, the blood is analysed directly. For a description of current GF-AAS methods for determining lead in blood the reader is referred to the literature (Christensen and Kristiansen 1993, D'Haese *et al.* 1991, Miller *et al.* 1987, Jacobson *et al.* 1991, Jin *et al.* 1990, Shuttler and Delves 1986, Angerer and Schaller 1985, Parson and Slavin 1993).

The description of a tested method for the determination of blood lead can be found in Volume 1 "Biological Monitoring of Chemical Exposure in the Workplace" (WHO 1996).

The determination of lead in blood must be carried out under conditions of statistical quality control. There are various commercially available control samples for internal quality control (WHO 1996). External quality control programmes which include the determination of blood lead in the occupational-medical and environmental-medical range are offered twice a year in Germany (Schaller *et al.* 1995, Lehnert *et al.* 1998, 1999).

6 Background Exposure

The internal exposure to lead in the general population has decreased considerably in recent years. At present the following reference values (95th percentiles) are given for the population of Germany by the Umweltbundesamt (Federal Environmental Agency):
– Women aged 25–69 years: 90 µg/l
– Men aged 25–69 years: 120 µg/l.
The median values are in the range from 30–40 µg/l (Ewers *et al.* 1999).

7 Evaluation of the BAT Value

The most sensitive parameter for the toxic effects of lead in men and women is the decrease in performance in behavioural tests. Effects on performance that can generally be reproduced begin at an average blood lead concentration of 400 µg/l. Other end points of lead toxicity, in particular effects on the peripheral nervous system and the kidneys, only become relevant in exposure ranges which are much higher.

It can therefore be assumed that a threshold limit value based on the avoidance of central nervous effects also offers protection against adverse effects on the peripheral nervous system and the kidney system, including potential nephrocarcinogenicity. Similar conclusions can therefore also be drawn for other toxic effects of lead, such as haem synthesis and effects on blood pressure.

There is considerable uncertainty about lead-induced adverse effects on reproduction in man. There is evidence that for men only blood lead concentrations above 400 µg/l may be associated with impairment of fertility.

Another consideration for the setting of threshold limit values at the workplace is the existence of background levels resulting from general environmental sources, and which are present also without occupational exposure (see "Kommission Human-Biomonitoring des Umweltbundesamtes" (Ewers *et al.* 1999)).

Under these circumstances the BAT value for women older than 45 years and for men has been changed to

400 µg lead per l blood

For women younger than 45 years the previous BAT value of 300 µg lead per litre blood has been reduced to 100 µg lead per litre blood (see Addendum).

The values must be reassessed if sufficient evidence should become available that behavioural effects may occur even at exposure levels in the range between 100 and 400 µg/l blood. The Commission recommends the BAT value for women older than 45 years and for men be set at 400 µg/l blood for the following reasons:

– There is evidence of beginning effects at a current blood lead range between 300 and 400 µg/l in 3 of the around 30 publications evaluated. It cannot be excluded that the effects may have been caused by earlier much higher levels of lead exposure. This will have to be borne in mind in the evaluation of further publications.
– The BAT value is conceived as a threshold limit value for the individual (ceiling value). On the basis of the existing data a value of 400 µg/l therefore seems justified.

8 Interpretation of Data

Occupational-medical health surveillance of persons exposed to lead and its inorganic compounds is carried out in Germany in accordance with guideline G2 issued by the Berufsgenossenschaft (Employers' Liability Insurance Association). Here the intervals between follow-up examinations and for additional biological monitoring are also given. With borderline findings more frequent monitoring may be necessary.

Because lead has a long half-life in the human organism, the blood samples may be taken at any time.

Abbreviations used in Tables 1a-d, 2a-c and 3:

PbB_{max}	=	maximum level of lead in blood
PbB	=	level of lead in blood
cumPb	=	cumulative level of lead in blood
AM	=	arithmetic mean
CBLI	=	cumulative blood lead index
SD	=	standard deviation
TWA-PbB	=	time-weighted average level of lead in blood
IBL	=	cumulative blood lead over the whole period of employment
ALA-D	=	δ-aminolaevulinic acid dehydratase
ALA-U	=	δ-aminolaevulinic acid in urine
ZPP	=	zinc protoporphyrin
MPb	=	lead in urine mobilized by calcium disodium ethylenediaminetetraacetate (CaEDTA)
EPI	=	Eysenck Personality Inventory

Blood lead values are given in µg/100 ml
The parameters for lead exposure have been rounded, mostly to two significant figures
N.B. "Tests/variables": these are generally procedures, also symptom and mood scales are given.

Lead and its compounds 53

Table 1a. Epidemiological studies of occupational exposure to lead included in the evaluation of a threshold limit value (cumulative exposure to lead)

Exposed group/ control group	Exposure variables	Tests/ variables	Significant differences between the groups	Significant correlations, dose/effect	Checks for confounders	Evaluation of the study
Haenninen et al. 1998						
exposed: 54 workers/ battery factories (43 men/11 women)		12/22	high versus low (PbB_{max}): • digit-symbol test • embedded figures • spatial memory • tiredness (**tests in reverse order**)	*in the total group*: with cumulative PbB (whole working life): • digit-symbol test • embedded figures • spatial memory • similarities with maximum PbB (whole working life): • digit-symbol test • embedded figures • spatial memory • similarities • Mosaic test with average PbB over the whole working life (is an indicator of current exposure here): • Santa Ana co-ordination (dominant hand) • digit-symbol *at low exposure levels*: with maximum PbB (whole working life): • embedded figures • symptoms (tiredness) • mood variables (tension, depression, helplessness) with cumulative PbB (whole working life): • mood variables (helplessness)	comparability of the groups with regard to: • working conditions • ethnic provenance • mother tongue • social status • job no diseases of the central nervous system no exposure to other neurotoxic substances no excessive alcohol consumption investigator unaware of level of exposure consideration of the influence of: • age • sex • education	satisfactory consideration of confounders

Table 1a. continued

Exposed group/ control group	Exposure variables	Tests/ variables	Significant differences between the groups	Significant correlations, dose/effect	Checks for confounders	Evaluation of the study
				with average PbB over the whole working life (is an indicator of current exposure here): • Mosaic test • similarities • mood variables (helplessness) with cumulative PbB in the last 3 years: • digit-symbol test • Mosaic test • spatial memory • memory span for digits *at high exposure levels:* with cumulative PbB (whole working life:) • embedded figures with maximum PbB (whole working life): • digit-symbol (memory)		
Chia *et al.* 1997						
exposed: 50 workers/ battery factory controls: 97 workers/ garage	• current PbB: exposed: 37/13–65 (geometric mean/range) controls: 6/2–12 (geometric mean/range) • cumPb (cumulative PbB, 1979–1997)	8/12	exposed versus controls: Santa Ana co-ordination (not dominant hand) grooved pegboard (2 variables) memory span for digits (forwards)	> 35 years of age: with current PbB: • trail making A (better with increasing lead level) with cumulative PbB: • digit-symbol • trail making A	controls had no exposure to neurotoxic substances no continuous medication exclusion criteria: • previous illness • regular alcohol consumption	on the whole satisfactory consideration of confounders

Table 1a. continued

Exposed group/ control group	Exposure variables	Tests/ variables	Significant differences between the groups	Significant correlations, dose/effect	Checks for confounders	Evaluation of the study
	exposed: 264/176/10–1146 (median/AM/range)		pursuit aiming (correct answers) trail making (A and B)		both groups unaware of purpose of the study investigator unaware of level of exposure same examination time for all persons (8–10 am) consideration of the influence of: • age • education • ethnic provenance • nicotine consumption • alcohol consumption	
Österberg 1997						
exposed: 38 workers/ lead smelting controls: 19 workers/ machine factory	• current PbB: low: 34/17–55 (median/range) high: 38/19–50 (median/range) controls: 4/1–7 (median/range) • CBLI (cumulative blood lead index)/μmol/l × month: low: 143/27–477 (median/range) high: 233/74–948	9/47	low/high versus control group; **controls worse!** • drawing figures (time for not dominant hand) • marking mid-points (time for not dominant hand)	with current PbB: • continuous graphic performance (errors) • acute symptoms (2 variables e.g. irritation) with CBLI: • Bourdon-Wiersma (omissions/errors) • remembering faces (errors) • activity (higher in controls) with maximum PbB: • Bourdon-Wiersma (errors) • long-term symptoms (e.g. concentration)	three parallel groups with regard to: • age • education • job comparability of the groups with regard to • alcohol consumption • nicotine consumption no relevant previous illnesses computer tests	consideration of all relevant confounders; it is questionable, however, whether groups are comparable enough with regard to premorbid intelligence (results not given)

Table 1a. continued

Exposed group/ control group	Exposure variables	Tests/ variables	Significant differences between the groups	Significant correlations, dose/effect	Checks for confounders	Evaluation of the study
	• max. PbB: low: 57/34–78 (median/range) high: 63/46–90 (median/range) • bone lead					
Lindgren et al. 1996						
exposed: 467 male workers/lead smelting	• current PbB: 28/8 (AM/SD) range ? • TWA PbB: 40/4–66 (AM/range) • IBL (cumulative blood lead over the whole period of employment): 765/0.6–1626 µg $\times y \times dl^{-1}$ (AM/range) low: 269/195–66 (AM/SD) moderate: 821/122–66 (AM/SD) high: 1228/145 (AM/SD)	9/14	low versus high: • digit-symbol test • logical memory (short-term) • trail making (2 variables) • Purdue pegboard (dominant hand) low versus moderate: not clear from the data	not investigated	no significant differences with regard to: • speech • slight neurological disturbances consideration of significant differences with regard to: • age • education • depression • alcohol consumption exclusion criteria: • severe neurological diseases • recent psychiatric illness consideration of the suppressor variable duration of employment investigator unaware of level of exposure investigations at the beginning of the shift	appropriate inclusion of all relevant confounders in the analysis

Table 1b. Epidemiological studies of occupational exposure to lead which fulfil the criteria for inclusion in the evaluation of a threshold limit value (repeated determination)

Exposed group/ control group	Exposure variables	Tests/ variables	Significant differences between the groups	Significant correlations, dose/effect	Checks for confounders	Evaluation of the study
Stollery et al. 1989 exposed: 86 workers/battery factory/printing industry	• current PbB: low: < 20 moderate: 21–40 high: 41–80 range 5–72 • ALA-U: group 1: 2.5 group 2: 4.0 group 3: 5.7 range 0.5–22 mg/l • ZPP	5/28	high versus moderate + low: • forming and remembering categories (identifying objects which belong in one class and those which do not, difficult choices) • choice reaction (number of attempts; decision time; speed of movement)	with current PbB: • forming and remembering categories (identifying objects which belong in one class and those which do not, difficult choices) • choice reaction (number of attempts; decision time; speed of movement)	consideration of the influence of: • exposure duration • age • age left school • sleep • alcohol consumption • job • stress/arousal comparability of the groups with regard to: • ethnic provenance • regional provenance computer tests	consideration of previous illnesses unclear consideration of the age on leaving school is possibly an inadequate way of checking the level of premorbid intelligence
Stollery et al. 1989 exposed: 70 workers	• average PbB at 3 investigation times within 8 months: low: 14/–20 (AM/range) moderate: 31/21–40 (AM/range) high: 52/41–80 (AM/range) • ALA-U • ZPP	5/28	high versus low: • logical thinking (reaction time reduced in group with high exposure, errors not reduced in high group) • forming and remembering categories (errors in learning)	with current PbB: • logical thinking (errors) • forming and remembering categories (errors in learning) • choice reaction (decision time; speed of movement, items skipped at high speeds) with ZPP: • forming and remembering categories with ALA-U: • forming and remembering categories	see above	see above

Table 1b. continued

Exposed group/control group	Exposure variables	Tests/variables	Significant differences between the groups	Significant correlations, dose/effect	Checks for confounders	Evaluation of the study
Yokoyama et al. 1998						
exposed: 19 workers/foundry controls: 12 workers/foundry (same company) after 2 years 17/10	• current PbB: exposed: 30–64 (range) controls: 8–20 (range) high: 40–64 low: 30–39 controls: 8–20 • current PbB after 2 years: high: 26–59 ** low: 24–39 controls: 8–14 • lead in urine (MPb)	5/5	Investigation 1: high versus low: • completing pictures high versus controls: • completing pictures Investigation 2: none found	in the exposed persons at the time of investigation 1: with current PbB: • completing pictures with MPb: • completing pictures with ALA-D: • completing pictures in the exposed persons at the time of investigation 2: not reported	comparability of the groups with regard to: • age • education • alcohol consumption no exposure to other neurotoxic substances no relevant previous illnesses no alcohol or drugs consumed on the day of the test	controls younger exposed group drinks more as the effect on performance disappears after the level of lead exposure is lowered both these points seem to be of little importance
Mantere et al. 1984						
exposed: 24 employees/battery factory after 1 year: 24 (6 women) after 2 years: 16 (6 women) after 4 years: 11 (5 women) controls: 33 workers (4 women)/cable production/power plant after 1 year: 33 after 2 years: 31	• current PbB: exposed: 16/7–30 (AM/SD) after 1 year: 31/13–49 (AM/SD) after 2 years: 31/14–46 (AM/SD) after 4 years: 29/17–45 (AM/SD) controls: 11/1–21 (AM/SD)	8/12	at beginning of exposure, exposed versus controls: • none after 1 year exposed versus controls: • Mosaic test (deterioration) • Santa Ana coordination (deterioration) after 2 years exposed versus controls: • Mosaic test (deterioration)	not investigated	exclusion criteria: • (other) neurotoxic exposure • previous illnesses comparability with regard to: • education • alcohol consumption • medicine consumption • nicotine consumption	consideration of confounders on the whole satisfactory (unclear about sex and job) exposed persons with higher level of performance in the initial test deteriorate less and/or have greater improvements in performance

Table 1b. continued

Exposed group/ control group	Exposure variables	Tests/ variables	Significant differences between the groups	Significant correlations, dose/effect	Checks for confounders	Evaluation of the study
after 4 years: 10 (4 women)	after 1 year: 11/6–18 (AM/SD) after 2 years: 11/5–21 (AM/SD) after 3 years: 7/4–12 (AM/SD) • TWA–PbB (1–4 years): exposed: constant 29 (AM) controls: 11 (AM) after 4 years: 8 (AM) low: ≤ 27 high: > 27 • ZPP		• Santa Ana coordination (deterioration) • memory span for digits (deterioration) after 4 years exposed versus controls: • determination of significance not carried out as number of persons too small at the beginning of exposure high versus controls/low versus controls: • none after 1 year high versus controls/low versus controls: • none after 2 years high versus controls: Mosaic test after 2 years low versus controls: Mosaic test Santa Ana coordination (**only in the low group**)		consideration of the influence of: • age • performance in initial test examiner unaware of level of exposure	

Table 1c. Epidemiological studies of occupational exposure to lead which fulfil the criteria for inclusion in the evaluation of a threshold limit value (TWA)

Exposed group/ control group	Exposure variables	Tests/ variables	Significant differences between the groups	Significant correlations, dose/effect	Checks for confounders	Evaluation of the study
Maizlish et al. 1995						
exposed: 43 workers/ lead smelting controls: 47 workers/ glass factory	• current PbB: exposed: 43/12/9–68 (AM/SD/range) controls: 15/6 (AM/SD) <10 4 % 10–25 44 % 25–39 20 % 40–61 26 % 61–81 6 % • maximum PbB (1986–1993): exposed: 60/20 (AM/SD) controls: 15/6 (AM/SD) • TWA-PbB: exposed: 48/12 (AM/SD) controls: 15/6 (AM/SD)	7/14	exposed versus controls: • symptoms reported in the year before the study (concentration, annoyance or agitation without reason, exceptional tiredness, painful joints)	with current PbB > PbB$_{max}$ & TWA with current PbB: • POMS (tension/anxiety, animosity, tiredness, depression) with maximum PbB: • POMS (animosity, depression) TWA PbB: • POMS (animosity, depression)	comparability of the groups with regard to: • job • geographical provenance consideration of the influence of: • age • education • alcohol consumption • earlier exposure to solvents • state of health examiner not 'blind'	• consideration of confounders satisfactory • as a result of the dismissal of many persons exposed long-term shortly before the study was carried out, it is questionable whether the group was representative (healthy worker effect?)
Parkinson et al. 1986						
exposed: 288 workers/ 3 battery factories controls: 181 workers/ chassis production	• current PbB: exposed: 40/13 (AM/SD) range ? controls: 7/3 (AM/SD) range ?	17/18	none	with current PbB: • psychosocial variable (conflicts) with ZPP: • psychosocial variable (annoyance at the workplace)	comparability of the groups with regard to: • ethnic provenance • psychiatric history • drug/alcohol abuse	• despite group comparison, the exposure levels for the controls were not given exactly (≤35) • adequate inclusion of all relevant confounders in the analysis

Table 1c. continued

Exposed group/ control group	Exposure variables	Tests/ variables	Significant differences between the groups	Significant correlations, dose/effect	Checks for confounders	Evaluation of the study
	• TWA–PbB: 49/12 (AM/SD) • maximum PbB: 79/29 (AM/SD) • PbB > 60: 0.23/0.22 (AM/SD) • ZPP			with TWA–PbB: • psychosocial variables (conflicts, 2 variables) • trait-annoyance • number of accidents at work • Mosaic test **(better with increasing exposure)** with maximum PbB: • psychosocial variables (conflicts, 2 variables, annoyance at the workplace) with PbB > 60: • psychosocial variables (conflicts, 2 variables) • number of accidents at work	consideration of the influence of: • age • education • income controls had no known exposure to neurotoxic substances exclusion criteria: • relevant neurological or other illnesses	
Ryan et al. 1987		18/21	exposed versus controls: • psychomotor velocity/manual dexterity factor • grooved pegboard (2 variables) • embedded figures (time, **in reverse order**)	not investigated		

Table 1c. continued

Exposed group/ control group	Exposure variables	Tests/ variables	Significant differences between the groups	Significant correlations, dose/effect	Checks for confounders	Evaluation of the study
Hogstedt et al. 1983 exposed: 49 workers who took part in occupational-medical health check-ups/lead smelting/battery factory controls: 27 industrial workers who took part in occupational-medical health check-ups/wire industry/turnery/munitions factory	• current PbB: exposed: 42 (AM) controls: 15 (AM) • TWA-PbB: exposed: 48 (AM) highest individual value 65 of 14 persons one had a value >69 range 27–69 high: >53 low: 27–52 controls: <21 • ZPP	15/7	low versus controls: • memory factor • learning factor high versus controls: • memory factor • simple reaction time high + low versus controls: • memory factor • learning factor (**here there were also decreases in performance from high to low**) • simple reaction time • irritability	significance unclear	comparability of the groups with regard to: • alcohol consumption • previous illnesses • exposure to solvents consideration of the influence of: • age only persons with the same level of schooling were allowed to take part similar examination times for all examiner aware whether the persons were exposed or belonged to the controls	• the lack of differences in verbal tests indicate comparable premorbid intelligence • consideration of confounders largely satisfactory • differences in the number of shift workers between the exposed and control groups may have led to underestimation of the effects, as shift workers generally had poorer performance

Table 1d. Epidemiological studies of occupational exposure to lead in the evaluation of a threshold limit value (single determination of current blood lead)

Exposed group/controls	Exposure variables	Tests/ variables	Significant differences between the groups	Significant correlations, dose/effect	Checks for confounders	Evaluation of the study
Pasternak et al. 1989						
exposed: 24 workers/ production of electrical components controls: 29 workers/same company	77 % exposure ≤ 3 years last 18 months without notable exposure • exposure years × median PbB: exposed: 128/156/19–680 (AM/SD/range) • PbB$_{max}$: exposed: 67/14/48–105 (AM/SD/range) • exposure years × median ZPP: exposed: 439/556/22–2008 (AM/SD/range) • ZPP$_{max}$: 265/136/39557 (AM/SD/range)	27/35	exposed versus controls: • tapping (2×) • memory span for digits (forwards) • visual memory • deletion test • dynamometer (only for the women) • anxiety • tiredness • confusion	• not investigated	exclusion criteria: • intake of psychoactive substances • previous illnesses comparability of the groups with regard to: • age • sex • ethnic provenance • family status • level of education/ premorbid intelligence • alcohol consumption • computer experience computer tests investigator unaware of level of exposure	• results of the vocabulary test indicate comparable premorbid intelligence • consideration of confounders largely satisfactory • no data for the exposure of the controls
Arnvig et al. 1980						
exposed: 9 workers/ battery factory	current PbB: 69/58–82 (mean/range)	17/22	• could not be investigated, as no controls available	• not investigated	average premorbid intelligence no relevant previous illnesses no excessive alcohol consumption both tests carried out at the same times for all in some cases consideration of the influence of: • age	• inclusion of the relevant confounders • inappropriately small number of samples • as a result of the exceptionally poor performance of those investigated and reported observations (impression of pre-senile dementia) it is questionable whether the group was representative

Table 2a. Epidemiological studies of occupational exposure to lead which are to be used with discretion in the evaluation of a threshold limit value (repeated determination)

Exposed group/ control group	Exposure variables	Tests/ variables	Significant differences between the groups	Significant correlations, dose/effect	Checks for confounders	Evaluation of the study
Dotzauer 1990						
exposed: 47 employees/ battery factory controls: 14 employees after 18 months: exposed: 11	• current PbB: controls: < 30 low: 40–60 high: > 60 range ? • ALA–U (μmol/100ml): controls: < 39 low: 76 high: > 76.4	4/11	high versus low + controls: • recognition of faces high + low versus controls: • choice reaction (errors/tempo self-determined) significant improvements after 18 months: • aiming • d2 • recognition of figures • choice reaction (hits/tempo self-determined)	with current PbB: • choice reaction (hits and errors/tempo self-determined, errors/ tempo predetermined) with ALA–U: • aiming (errors) • choice reaction (errors/tempo self-determined, errors/ tempopredetermined)	comparability of the groups with regard to: • age • education investigator unaware of level of exposure	• alcohol consumption and job not considered as confounders • previous illnesses not taken into consideration • it is not possible to check whether the consideration of other confounders was appropriate • no data for the exposure at the second investigation
Baker et al. 1983/1984/1985						
exposed: 99 workers/ foundry controls: 61 workers/ assembly line in the second year of investigation: 43/34 in the third year of investigation: 38/19	in the first year of investigation: current PbB: 0–20 16 % 21–40 61 % 41–60 18 % 61–80 5 % range 3–80 • TWA PbB (12 months)	15/23	in the first year of investigation: significance not investigated	in the first year of investigation: with current PbB: • vocabulary test • finding similarities • mental control • POMS (annoyance, depression, tiredness, confusion) with TWA–PbB: • continuous performance	comparability of the groups with regard to: • ethnic provenance • job • wage consideration of the influence of: • education • age • sex	• no comments on the significance of group differences in the first year of investigation

Table 2a. continued

Exposed group/ control group	Exposure variables	Tests/ variables	Significant differences between the groups	Significant correlations, dose/effect	Checks for confounders	Evaluation of the study
	in the second year of investigation: • current PbB: range 7–59 in the third year of investigation: • current PbB: range 8–50 • maximum PbB (15 months) • TWA–PbB (15 months) • quadratic mean • TWA min > 40 • ZPP			• POMS (annoyance, depression, tiredness, confusion) in the second year of investigation: with TWA/quadratic mean/TWA min >40 > current PbB/max PbB with current PbB: • pair-associated learning with TWA–PbB: • pair-associated learning • finding similarities • vocabulary test • Santa Ana (both hands) • POMS (annoyance, depression, vitality, confusion) with quadratic mean: • pair-associated learning • Santa Ana (both hands) • POMS (annoyance, depression, vitality, confusion) with maximum PbB: • pair-associated learning • POMS (annoyance) with TWA min >40: • pair-associated learning • Santa Ana (both hands) • POMS (annoyance, depression, confusion)	for mood variables consideration of the influence of: • job exclusion criteria: • increased alcohol consumption • drug abuse • exposure to solvents • previous illnesses double blind study	• the correlations with the vocabulary test makes it seem questionable that confounders were adequately considered for premorbid intelligence. It is recognizable from the results, however, that the difference is not large

Table 2b. Epidemiological studies of occupational exposure to lead which are to be used with discretion in the evaluation of a threshold limit value (TWA).

Exposed group/ control group	Exposure variables	Tests/ variables	Significant differences between the groups	Significant correlations, dose/effect	Checks for confounders	Evaluation of the study
Braun & Daigneault 1991						
exposed: 41 workers/ lead smelting (30 no longer exposed for up to 84 months) controls: 37 workers/ printing, copying service	• maximum PbB: exposed: 87/22/49–158 (mean/SD/range) TWA–PbB: exposed: 53/8/30–66 (mean/SD/range)	12/23	exposed versus controls: • simple reaction time (scatter of the reaction time/length of the reaction time) • grip strength (exposed group better)	with TWA–PbB: • tapping (not dominant hand) • tapping (right foot) • grip strength (dominant/ not dominant hand) • grooved pegboard (dominant/not dominant hand) • motor performance factor with maximum PbB: • tapping (not dominant hand) • grooved pegboard (dominant/not dominant hand)	exclusion criteria: • poor performance in a vocabulary test • ethnic provenance • neurological/ psychiatric and other relevant previous illnesses • exposure to other heavy metals comparability of the groups with regard to: • age • years of schooling • alcohol consumption • other chemical exposures • calculation test • body weight • height	• the mixing of persons no longer exposed for 7 years with those currently exposed does not seem sensible in view of the uncertainty about the reversibility of lead-induced effects • the analysis of other toxic exposures indicates that the results may be confounded by other exposures (excluded only for controls)
Haenninen et al. 1978						
exposed: 49 workers/ battery factory/ machine factory (10 women) controls: 24 workers/ battery factory/ electronics factory	• current PbB: exposed: 32/11 (AM/SD) controls: 12/4 (AM/SD) range ? • maximum PbB: exposed: 50/12(AM/SD)	13/29	exposed versus controls: • tiredness • sleep disturbances • absence • depression • lability • neurovegetative symptoms • gastrointestinal complaints	with TWA > maximum PbB > current PbB with current PbB: • repetition of spoken numbers • Santa Ana co-ordination (right hand) with TWA–PbB: • Mosaic test • visual reproduction	comparability of the groups with regard to: • sex • education • job exposed persons tended to be younger neuroticism taken into account	• inadequate consideration of confounders (age distribution) • alcohol consumption and previous illnesses not considered as confounders

Table 2b. continued

Exposed group/ control group	Exposure variables	Tests/ variables	Significant differences between the groups	Significant correlations, dose/effect	Checks for confounders	Evaluation of the study
	• TWA–PbB: exposed: 38/9 (AM/SD)		• neurological symptoms I • neurological symptoms II • neuroticism high versus low (only for symptoms and EPI): • absence • gastrointestinal symptoms • neurological symptoms II	• Santa Ana co-ordination (right, left, both hands) with maximum PbB: • Santa Ana co-ordination (right, left, both hands)		

Table 2c. Epidemiological studies of occupational exposure to lead which are to be used with discretion in the evaluation of a threshold limit value (single determinations of current blood lead level)

Exposed group/ control group	Exposure variables	Tests/ variables	Significant differences between the groups	Significant correlations, dose/effect	Checks for confounders	Evaluation of the study
Jeyaratnam et al. 1986						
exposed: 49 workers/ lead processing controls: 36 workers	• current PbB (AM/SD): exposed: 49/15 controls: 15/3 range ?	9/18	exposed versus controls: • digit-symbol test • Bourdon-Wiersma (speed) • trail-making (A) • Santa Ana co-ordination (right, left hand) • flicker fusion frequency • simple reaction time • subjective symptoms (anxiety, depression, lack of concentration, forgetfulness, headaches, dizziness)	not investigated	level of education comparable significant differences with regard to: • age	• inadequate consideration of confounders (controls older) makes the results seem more meaningful • alcohol consumption, job and previous illnesses not considered as confounders • exposure only estimated for controls
Williamson & Teo 1986						
exposed: 59 male workers/battery factory, lead smelting controls: 49 men, 10 women	• current PbB: exposed: 49/13/ 25–82 (median/ SD/range) controls: ? • current PbB also as the mean value in 3 years	9/20	exposed vs. controls: • flicker fusion frequency • simple reaction time • visual pursuit (slower at both speeds) • hand steadiness (2 variables) • sensory memory (3 variables) • Sternberg short-term memory (2 speed variables)	with current PbB: • sensory memory (1 variable)	no significant differences with regard to: • age • type of job • duration of employment • level of education • cigarette consumption • alcohol consumption no toxic exposure of the controls	• questionable whether consideration of the confounder age was adequate (despite a lack of significance, median of the controls 4.5 years lower) • previous illnesses not considered as confounders • data for the exposure of the controls not given

Table 2c. continued

Exposed group/ control group	Exposure variables	Tests/ variables	Significant differences between the groups	Significant correlations, dose/effect	Checks for confounders	Evaluation of the study
			• pair-associated learning/short-term reproduction (number of reproductions, number of attempts, success)			• groups not comparable with regard to sex

Campara et al. 1984/Zimmermann-Tansella et al. 1983

Exposed group/ control group	Exposure variables	Tests/ variables	Significant differences between the groups	Significant correlations, dose/effect	Checks for confounders	Evaluation of the study
exposed: 40 workers/ battery factory controls: 20 male nurses	• current PbB: controls: 20/6/11–27 (AM/SD/range) low: 32/3/26–35 (AM/SD/range) high: 52/5/46–60 (AM/SD/range) • maximum PbB: low: < 40 high: < 60	16/34	high versus low: • finding similarities • vocabulary test • digit-symbol test • completing pictures • deletion test (2 error variables) • Rey Osterrieth • SSQ22 (in particular motor disturbances) • general performance • logical thinking high versus controls: • copying symbols • finding similarities • vocabulary test • digit-symbol test • completing pictures • deletion test (2 error variables) • Rey Osterrieth • SSQ22 (in particular motor disturbances, paraesthesia and weakness in the extremities) low versus controls: • motor disturbances	with current PbB: • finding similarities • vocabulary test • digit-symbol test • completing pictures • sorting cards (decision time) • general performance • logical thinking • SSQ20 (physical symptoms) • group of gastrointestinal symptoms • group of neurological symptoms	comparability of the groups with regard to: • family status • distance to workplace • chronic illnesses • duration of employment • age • schooling • neuroticism/extraversion no participants with • increased alcohol consumption • drug abuse in the last 6 months • relevant previous neurological or psychiatric illnesses investigator unaware of level of exposure same examination time	• job not considered as a confounder • the relationship between the vocabulary test results and blood lead concentration indicates that the different kinds of jobs are also linked with a different premorbid intelligence and the results could be confounded by this

Table 2c. continued

Exposed group/ control group	Exposure variables	Tests/ variables	Significant differences between the groups	Significant correlations, dose/effect	Checks for confounders	Evaluation of the study
Valciukas et al. 1980						
exposed: 141 workers/ cable processing New York/cable production New York/lead smelting California controls: 265 persons from Michigan who took part in health check-ups	• current PbB (AM/SD): exposed: cable processing 28/6 cable production 39/11 lead smelting 51/12 range: < 30– > 80 • ZPP	3/4	not carried out	with ZPP > current PbB: with current PbB: • Mosaic test (in the group of lead smelters) • summarized performance index (in the group of lead smelters) with ZPP: • embedded figures (in the group of cable producers) • summarized performance index (in the group of cable producers) • Mosaic test (in the group of lead smelters) • digit-symbol test (in the group of lead smelters) • embedded figures (in the group of lead smelters) • summarized performance index (in the group of lead smelters)	consideration of the influence of: • age • education investigator unaware of level of exposure	• lack of current PbB for controls less important as no group comparison was carried out • alcohol consumption, job and previous illnesses not considered as confounders
Spivey et al. 1979						
exposed: 69 workers/ lead smelting, processing controls: 35 workers/ aluminium smelting, processing	• current PbB: exposed: 61/13 (AM/SD) controls: 22/2 (AM/SD) • ALA-D	1/3	exposed versus controls: • neurological symptoms (difficulties in calculating, dejection, dizziness,	with current PbB: • total number of symptoms with ALA-D: • total number of symptoms	exclusion criteria: duration of employment < 1 year comparability of the groups with regard to: • job	• very careful selection of the groups with regard to exposure conditions/ job

Lead and its compounds 71

Table 2c. continued

Exposed group/ control group	Exposure variables	Tests/ variables	Significant differences between the groups	Significant correlations, dose/effect	Checks for confounders	Evaluation of the study
			concentration difficulties, wobbly handwriting, weakness in the hands) • other symptoms (painful joints, metallic taste, nocturia, racing heartbeat, watery eyes (not allergic))		• previous illnesses • alcohol consumption • nicotine consumption importance of differences unclear for: • age • schooling significant differences with regard to: • exposure to arsenic and cadmium (is presented as being in the normal range) consideration of the influence of: • ethnic provenance • arsenic • schooling	• positive is that other toxic exposures were checked; it is unclear whether differences are influenced by the level of premorbid intelligence • that the symptoms decrease with increasing duration of employment could also indicate that habituation/compensation effects occur
Grandjean et al. 1978						
exposed: 42 workers/ battery factory/lead processing/radiator repair/cable production controls: 22 workers/ oil mill	• current PbB: exposed: 46/13–88 (median/range) controls: 17/11–27 (median/range) • ZPP • lead in hair	14/30	exposed versus controls (2-sided test!): • visual production/ reproduction **(exposed better)** • continuous graphic performance (time) • tapping (not dominant hand) • number learning **(exposed better)**	(only within the exposed group) with current PbB: • visual production (learning) • number learning • Wechsler intelligence test (general understanding, arithmetic, finding similarities, memory span for digits, digit-symbol, picture completion, Mosaic test, verbal IQ, performance IQ, total IQ)	comparability of the groups with regard to: • education • sex the influence of age taken into account in the Wechsler intelligence test no participants with relevant previous neurological or psychiatric illnesses	• significant difference in premorbid intelligence (vocabulary test) in the exposed and control groups • inadequate consideration of the confounder alcohol consumption is of less importance as the exposed group drink less

Table 2c. continued

Exposed group/ control group	Exposure variables	Tests/ variables	Significant differences between the groups	Significant correlations, dose/effect	Checks for confounders	Evaluation of the study
			• Wechsler intelligence test (general knowledge, general understanding, arithmetic, finding similarities, vocabulary, digit-symbol, Mosaic test, ordering pictures, verbal IQ, performance IQ, total IQ)	with ZPP: • visual production (learning) • continuous graphic performance (errors, time) • Wechsler intelligence test (general understanding, finding similarities, memory span for digits, digit-symbol, Mosaic test, verbal IQ, performance IQ, total IQ)	different number of slight head injuries alcohol consumption lower in the exposed group examiner aware whether the persons were exposed or belonged to the controls	
Valciukas et al. 1978						
exposed: 90 workers/ lead smelting (1 woman) controls: 25 workers/ steel works 99 workers/paper factory 93 farmers	• current PbB: exposed: <40 17 % 40–59 61 % 60–79 20 % >80 1 % controls: <40 94 % 40–59 6 % range: <30– >80 • ZPP	4/5	exposed versus controls: • Mosaic test • digit-symbol test • embedded figures	with ZPP > PbB: with ZPP: • Mosaic test • digit-symbol test • embedded figures with current PbB: • Mosaic test • embedded figures with exposure duration: • digit-symbol test • embedded figures • Santa Ana co-ordination (both hands)	comparability of the groups with regard to: • sex • ethnic provenance • job • life-style consideration of the influence of: • education • age investigator unaware of level of exposure exclusion criteria controls: good state of health no exposure to neurotoxic substances	• alcohol consumption and previous illnesses not considered as confounders

Table 3. Epidemiological studies of occupational exposure to lead which are not of use for the evaluation of a threshold limit value

Exposed group/ control group	Exposure variables	Tests/ variables	Significant differences between the groups	Significant correlations, dose/effect	Checks for confounders	Evaluation of the study
Banks & Stollery 1988						
exposed: 40 workers/ printing industry controls: ? workers/ printing industry (same company)	• current PbB: exposed: 31 (mean?) controls: 15 (mean?)	1/3	exposed versus controls in the current comparison: • logical thinking (errors) exposed versus controls in the course of the year: • logical thinking (errors decrease in the exposed, similar to the controls) • logical thinking (reaction time dependant on difficulty of the task, no improvement in the exposed)	with current PbB (only in the exposed group, with the blood lead level which changed over the course of the year): • logical thinking (errors in complex tasks)	comparability of the groups with regard to: • age on leaving school • sleep • stress comparability questionable as not analysed for: • age • alcohol consumption • arousal • waking time	• previous illnesses and job not considered as confounders • age on leaving school is not a reliable indicator of the level of premorbid intelligence • although waking time and alcohol consumption differ greatly in the exposed and control groups and correlate significantly/highly significantly with the errors, it is not recognizable that they were included in the analysis as confounders; it cannot be excluded that the effects have to be attributed to this • the levels of exposure given provide only an inexact impression

Table 3. continued

Exposed group/ control group	Exposure variables	Tests/ variables	Significant differences between the groups	Significant correlations, dose/effect	Checks for confounders	Evaluation of the study
Repko et al. 1978 exposed: 85 workers/ battery factory (81 men, 4 women) controls: 55 workers/ production of illuminants, repair workshops, unemployed persons (46 men/9 women)	• current PbB: exposed: 46 controls: 18 range ? • ALA-D: exposed: 27 units/l red blood cells controls: 23 units/l red blood cells • ALA-U: exposed: 0.7 controls: 0.3 • lead in urine • free erythrocyte protoporphyrin	6/53	exposed versus controls: • eye-hand co-ordination (1 variable) • simple reaction time (1 variable) • pure tone auditory thresholds (7 variables/ 6 frequencies) • tone decay (7 variables) • enjoyment-disgust, **high in controls!** • annoyance, **higher in controls!** • accidents at work	with ALA-U: • pure tone auditory thresholds (1 variable) • tone decay (1 variable) with ALA-D: • tone decay (2 variables)	exclusion criteria: • lead exposure • exposure to toxic substances comparability of the groups with regard to: • age • sex • ethnic provenance • lack of inclination to answer in the sense of a social desirability in some cases consideration of the influence of: • sex group differences with regard to education	• as a result of the use of an unsuitable test, the extent of the difference in the level of education in the groups was underestimated • different confounders (job, alcohol consumption, previous illnesses) were not considered • positive is that inclination was checked, in the sense of a social desirability to answer, a tendency that could not be detected

9 References

Angerer J, Schaller KH (Eds) (1985) Lead in blood. in: Analyses of hazardous substances in biological materials, Vol. 1, VCH, Weinheim

Angerer J, Schaller KH (Eds) (1988) Lead in blood and urine. in: Analyses of hazardous substances in biological materials, Vol. 2, VCH, Weinheim

Angerer J, Schaller KH (Eds) (1999) Lead, ICP-MS collective method. in: Analyses of hazardous substances in biological materials, Vol. 6, Wiley-VCH, Weinheim

Araki S, Murata K, Aono H (1987) Central and peripheral nervous system dysfunction in workers exposed to lead, zinc and copper. Int Arch Occup Environ Health 59: 177–187

Arnvig E, Grandjean P, Beckmann J (1980) Neurotoxic effects of heavy lead exposure determined with psychological tests. Toxicol Lett 5: 399–404

Baker EL, Feldman R, White R, Harley J (1983) The role of occupational lead exposure in the genesis of psychiatric and behavioral disturbances. Acta Psychiatr Scand 67: 38–48

Baker EL, Feldman RG, White RA, Harley JP, Niles CA, Dinse GE, Berkey CS (1984) Occupational lead neurotoxicity: a behavioural and electrophysiological evaluation-study design and year one results. Br J Ind Med 41: 352–361

Baker EL, White RA, Pothier LJ, Berkey CS, Dinse GE, Travers PH (1985) Occupational lead neurotoxicity: Improvement in behavioral effects after reduction of exposure. Br J Ind Med 42: 507–516

Balbus-Kornfeld JM, Stewart W, Bolla KI, Schwartz BS (1995) Cumulative exposure to inorganic lead and neurobehavioural test performance in adults: an epidemiological review. Occup Environ Med 52: 2–12

Baloh RW (1974) Laboratory diagnosis of increased lead absorption. Arch Environ Health 28: 198–208

Banks HA, Stollery BT (1988) The longitudinal evaluation of verbal-reasoning in lead workers. Sci Total Environ 71: 469–476

Braun CMJ, Daigneault S (1991) Sparing of cognitive executive functions and impairment of motor functions after industrial exposure to lead: A field study with control group. Neuropsychol 5: 179–193

Campara P, D'Andrea F, Micciolo R, Savonitto C, Tansella M, Zimmerman-Tansella C (1984) Psychological performance of workers with blood-lead concentration below the current threshold limit value. Int Arch Occup Environ Health 53: 233–246

Chamberlain AC (1985) Prediction of response of blood lead to airborne and dietary lead from volunteer experiments with lead isotopes. Proceedings of the Royal Society B 224: 149–182

Chia SE, Chua LH, Ng TP, Foo SC, Jeyaratnam J (1994) Postural stability of workers exposed to lead. Occup Environ Med 51: 768–771

Chia SE, Chia HP, Ong CN, Jeyaratnam J (1996a) Cumulative blood lead levels and nerve conduction parameters. Occup Med 46: 59–64

Chia SE, Chia K-S, Chia HP, Ong CN, Jeyaratnam J (1996b) Three-year follow-up of serial nerve conduction among lead-exposed workers. Scand J Work Environ Health 22: 374–380

Chia SE, Chia HP, Ong CN, Jeyaratnam J (1996c) Cumulative concentrations of blood lead and postural stability. Occup Environ Med 53: 264–268

Chia SE, Chia HP, Ong CN, Jeyaratnam J (1997) Cumulative blood lead levels and neurobehavioral test performance. Neurotoxicology 18: 793–804

Christensen JM, Kristiansen J (1993) Lead. in: Handbook on metals in clinical chemistry. Marcel Dekker Inc., New York

D'Haese PCD, Lamberts LV, Liang L, Van de Vyver FL, De Broe ME (1991) Elimination of matrix and spectral interferences in the measurement of lead and cadmium in urine and blood by ET-AAS with deuterium background correction. Clin Chem 1991; 37: 1583–1588

Davis JM, Svendsgaard DJ (1990) Nerve conduction velocity and lead: A critical review and meta-analysis, in: BL Johnson (Ed.) Advances in neurobehavioral toxicology: Applications in environmental and occupational health. Lewis Publishers, Chelsea, Mi 373–376

Delves HAT, Campbell MJ (1988) Measurements of total lead concentrations and of lead isotope ratios in whole blood by use of inductively coupled plasma source mass spectrometry. J Anal Atom Spectr 3: 343–348

DFG (Deutsche Forschungsgemeinschaft) Seeber A, Bolt HM, Gelbke HP, Miksche L, Pawlik K, Rüdiger HW, Triebig G, Ziegler-Skylakakis K (1997) Verhaltenstoxikologie und MAK-Grenzwertfestlegungen. Wissenschaftliche Arbeitspapiere. Wiley-VCH Verlag, Weinheim

Dotzauer H (1990) Bleiexposition und psychische Leistungen – Ein Diskussionsbeitrag zu Dosis-Wirkungsbeziehungen nach kurzer Expositionsdauer. Z ges Hyg 36: 271–273

Ehle AL (1986) Lead neuropathy and electrophysiological studies in low level lead exposure – a critical review. Neurotoxicology 7: 203–216

Ewers U, Krause C, Schulz C, Wilhelm M (1999) Reference values and human biological monitoring values for environmental toxins. Int Arch Occup Environ Health 72: 255–260

Fricke R, Treinis G (1985) Einführung in die Metaanalyse. Huber, Bern

Fujimura Y, Araki S, Murata L, Sakai T (1998) Assessment of peripheral, central and autonomic nervous system functions in two lead smelters with high blood lead concentrations: A follow-up study. J Occup Health 40: 9–15

Gercken B, Barnes RM (1991) Determination of lead and other trace element species in blood by size exclusion chromatography and inductively coupled plasma/mass spectrometry. Anal Chem 63: 283–287

Grandjean P, Arnvig E, Beckmann J (1978) Psychological dysfunctions in lead-exposed workers – relation to biological parameters of exposure. Scand J Work Environ Health 4: 295–303

Greim H (Ed.) (2002) Lead and its inorganic compounds, apart from lead arsenate and lead chromate. in: Occupational toxicants, Vol. 17, Wiley-VCH, Weinheim

Haas Th, Wieck AG, Schaller KH, Mache K, Valentin H (1972) Die usuelle Bleibelastung bei Neugeborenen und ihren Müttern. Zbl Bakt Hyg I Abt Orig B 155: 341–349

Haenninen H, Aitio A, Kovala T, Luukhouen R, Matihainen E, Maumela T, Erkkitä J, Riihimäki V (1998) Occupational exposure to lead and neuropsychological dysfunction. Occup Environ Med 55: 202–209

Haenninen H, Hernberg S, Mantere P, Vesanto R, Jalkanen M (1978) Psychological performance of subjects with low exposure to lead. JOM: 683–689

Henschler D (Ed.) (1977) Blei und seine Verbindungen (berechnet als Pb) außer Bleiarsenat, Bleichromat und Alkylbleiverbindungen. Gesundheitsschädliche Arbeitsstoffe, Toxikologisch-arbeitsmedizinische Begründungen von MAK-Werten, 5. Lieferung, Verlag Chemie, Weinheim

Hogstedt C, Hane M, Bodin A, Agrell L (1983) Neuropsychological test results and symptoms among workers with well-defined long-term exposure to lead. Br J Ind Med 40: 99–105

Jacobson BE, Lockitch G, Quigley G (1991) Improved sample preparation for accurate determination of low concentrations of lead in whole blood by graphite furnace analysis. Clin Chem 37: 515–9

Jeyaratnam J, Boey K, Ong C, Chia C, Phoon W (1986) Neuropsychological studies on lead workers in Singapore. Br J Ind Med 43: 626–629

Jin Z, Shougui J, Shikum C, Desen J, Chahvalarle D (1990) Direct determination of lead in human blood and selenium cadmium, copper, zinc in serum by electrothermal atomic absorption spectrometry using Zeeman effect background correction. Fresenius J Anal Chem 337: 877–881

Kovala T, Matikainen E, Mannelin T, Erkkilä J, Riihimäki V, Hänninen H, Aitio A (1997) Effects of low level exposure to lead on neurophysiological functions among lead battery workers. Occup Environ Med 54: 487–493

Krigman M, Bouldin TW, Mushak P (1980) Lead. in: Spencer PS, Schaumburg HH (Eds) Experimental and clinical neurotoxicology. Williams & Wilins, Baltimore, 490–507

Lehnert G, Angerer J, Schaller KH (1998) Statusbericht über die externe Qualitätssicherung arbeits- und umweltmedizinisch-toxikologischer Analysen in biologischen Materialien. Arbeitsmed Sozialmed Umweltmed 33: 21–26

Lehnert G, Schaller KH, Angerer J (1999) Report on the status of the external quality-control programs for occupational-medical toxicological analyses in biologicals in Germany. Int Arch Occup Environ Health 72: 60–64

Lezak MD (1995) (Ed.) Neuropsychological assessment. New York: Oxford University Press, 3rd edition

Lindgren K, Masten V, Ford D, Bleecker M (1996) Relation of cumulative exposure to inorganic lead and neuropsychological test performance. Occup Environ Med 53: 472–477

Maizlish N, Parra G, Feo O (1995) Neurobehavioral evaluation of Venezuelan workers exposed to inorganic lead. Occup Environ Med 52: 408–414

Mantere P, Haenninen H, Hernberg S, Luukkonen R (1984) A prospective follow-up study on psychological effects in workers exposed to low levels of lead. Scand J Work Environ Health 10: 43–50

Meyer-Baron M, Seeber A (2000) A meta-analysis for neurobehavioural results due to occupational lead exposure with blood lead concentrations < 70 µg/100 ml. Arch Toxicol 73: 510–518

Miller DT, Paschal DC, Gunter EW, Strand PE, D'Angelo J (1987) Determination of lead in blood using electrothermal atomization absorption spectrometry with L'Vov platform and matrix modifier. Analyst 112: 1701–4

Murata K, Araki S, Aono H (1987) Effects of lead, zinc and copper absorption on peripheral nerve conduction in metal workers. Int Arch Occup Environ Health 59: 11–20

Murata K, Araki S (1991) Autonomic nervous system dysfunction in workers exposed to lead, zinc, and copper in relation to peripheral nerve conduction: a study of R–R interval variability. Am J Ind Med 20: 663–671

Murata K, Araki S, Yokoyama K, Uchida E, Fujimura Y (1993) Assessment of central, peripheral, and autonomic nervous system functions in lead workers: neuroelectrophysiological studies. Environ Res 61: 323–336

Ostapczuk P (1992) Direct determination of cadmium and lead in whole blood by potentiometric stripping analysis. Clin Chem 38: 1995–2001

Österberg K (1997) A neurobehavioural study of long-term occupational inorganic lead exposure. Sci Tot Environ 201: 39–51

Parkinson D, Ryan C, Bromet E, Connell M (1986) A psychiatric epidemiologic study of occupational lead exposure. Am J Epidemiol 123: 261–269

Parson PJ, Slavin W (1993) A rapid Zeeman graphite furnace atomic absorption spectrometric method for the determination of lead in blood. Spectrochim Acta 48B: 925–39

Pasternak G, Becker CE, Lash A, Bowler R, Estrim W, Law D (1989) Cross-sectional neurotoxicology study of lead-exposed cohort. Clin Toxicol 27: 37–51

Repko JD, Corum CR, Garcia PD, Jones LS (1978) The effects of inorganic lead on behavioural and neurologic function (US Dept of Health, Education, and Welfare, National Institute for Occupational Safety and Health). Cincinnatti: DHEW (NIOSH)

Roels H, Hubermont G, Buchet JP, Lauwerys R (1978) Placental transfer of lead, mercury, cadmium, and monoxide in women. III. Factors influencing the accumulation of heavy metals in the placenta and the relationship between metal concentration in the placenta and in maternal and cord blood. Environ Res 16: 236–247

Ryan CM, Morrow L, Parkinson D, Bromet E (1987) Low level lead exposure and neuropsychological functioning in blue collar males. Int J Neurosci 36: 29–39

Schaller KH, Angerer J, Lehnert G (1995) Current status of the external quality assurance programmes of the German Society for Occupational and Environmental Medicine. Toxicol Lett 77: 213–217

Seppäläinen AM, Hernberg S, Vesanto R, Kock B (1983) Early neurotoxic effects of occupational lead exposure: a prospective study. Neurotoxicol 4: 181–192

Shuttler IL, Delves HAT (1986) Determination of lead in blood by atomic absorption spectrometry with electrothermal atomization. Analyst 111: 651–60

Skerfving S (1993) Inorganic lead. in: Beije B, Lundberg P (Eds) Criteria documents from the Nordic expert group 1992. Arbeite och Hälsa 1: 125–238

Spivey G, Brown CP, Baloh RW (1979) Subclinical effects of chronic increased lead absorption – a prospective study. J Occup Med 21: 423–429

Spivey GH, Baloh RW, Brown P, Browdy BL, Campion DS, Valentine JL, Morgan DE, Culver BD (1980) Subclinical effects of chronic increased lead absorption – a prospective study. J Occup Med 22: 60–612

Stollery B, Banks H, Broadbent D, Lee W (1989) Cognitive functioning in lead workers. Br J Ind Med 46: 698–707

Teruya K, Sakurai H, Omae K, Higashi T, Muto T, Kaneko Y (1991) Effect of lead on cardiac parasympathetic function. Int Arch Occup Environ Health 62: 549–553

Task Group on Metal Accumulation TGMA (1973). Environ Physiol Biochem 3: 65–107

Triebig G (1984) Schwere Blei-Polyneuropathie. Fortschr Med 102: 485–488

US Environmental Protection Agency. (1986) Air quality criteria for lead. EPA–600/8–83/028aF, Vols I–IV, Environmental Protection Agency, Environmental Criteria and Assessment Office, Research Triangle Park, N.C

Valciukas JA, Lilis R, Eisinger J, Blumberg WE, Fischbein A, Selikoff IJ (1978) Behavioural indicators of lead neurotoxicity: results of a clinical field survey. Int Arch Occup Environ Health 41: 217–236

Valciukas JA, Lilis R, Singer R, Fischbein A, Anderson HA, Glickman L (1980) Lead exposure and behavioral changes: Comparisons of four occupational groups with different levels of lead absorption. Am J Ind Med 1: 421

WHO (World Health Organization) (1995) Inorganic lead. IPCS (International Programme on Chemical Safety). Environmental Health Criteria 165, WHO, Geneva

WHO (1996) Biological monitoring of chemical exposure in the workplace, Guidelines, Volume 1. WHO – HPR – OCH 1996.1, WHO, Geneva

Williamson A, Teo R (1986) Neurobehavioral effects of occupational exposure to lead. Br J Ind Med 43: 374–380

Yokoyama K, Araki S, Murata K, Morita Y, Katsuno N, Tanigawa T, Mori N, Yokata J, Ito A, Sakata E (1997) Subclinical vestibulo-cerebellar, anterior cerebellar lobe and spinocerebellar effects in lead workers in relation to concurrent and past exposure. Neurotoxicol 18: 371–380

Yokoyama K, Araki S, Aono H, Murata K (1998) Calcium disodium ethylenediaminetetraacetate-chelated lead as a predictor for subclinical lead neurotoxicity: follow-up study on gun-metal foundry workers. Int Arch Occup Environ Health 71: 459–464

Zimmerman-Tansella C, Campara P, D'Andrea F, Savonitto C, Tansella M (1983) Psychological and physical complaints of subjects with low exposure to lead. Human Toxicology 2: 615–23

Authors: H.M. Bolt, K.H. Schaller
Approved by the Working Group: 19.01.2000

Lead and its compounds, Addendum (except lead arsenate, lead chromate and alkyllead compounds)

BAT for women younger than 45 years   100 µg lead/l blood

Date of evaluation   2003

10 Re-evaluation of the BAT Value for Lead for Women below 45 Years of Age

More recent studies of the toxic effects of lead on reproduction in women and a reduction in the exposure to lead of the general population have made necessary the re-evaluation of the previous BAT value of 300 µg lead per litre blood for women younger than 45 years.

10.1 Exposure and effects

The current data for the developmental toxicity of lead can be found in the supplement to the MAK documentation for lead and its inorganic compounds, apart from lead arsenate and lead chromate (Greim 2002). Discussed there are the results of prospective studies of children (longitudinal studies) which determined the relationship between blood levels in prenatal or perinatal maternal blood or in umbilical cord blood and data for postnatal cognitive development. Seven birth cohorts were evaluated, which are presented in the MAK documentation (see Table 1 at the end of this addendum). Cross-sectional studies, which are primarily of environmental relevance, were not taken into consideration.

The seven studies evaluated are summarized in the MAK documentation (Greim 2002) as follows:

The meta-analyses of the prospective studies which have been carried out, which analysed only postnatal blood lead levels, can contribute only little to answering our question of the significance of prenatal and perinatal exposure to lead on later cognitive development (Pocock *et al.* 1994, WHO 1995).

"In spite of intensive efforts to standardize the methods used (body burden, measurement of effects, relevant confounders), the prospective studies are so different that a truly concurrent picture of the effects of lead cannot really be expected (WHO 1995). The various cohorts differ markedly in the correlates reflecting opportunities for cognitive development. The average maternal IQ in the Boston study cohort was, for example, 121, that in the Cleveland study cohort only 75; this reflects enormous socio-economic and familial differences. The nature of the exposure to lead also differs markedly in the various studies. It includes exposure via paint containing lead in dilapidated slum areas (Cincinnati), exposure mainly via traffic in cities like Sydney and Boston, exposure in the drinking water in one city (Glasgow), and emissions from lead smelters (Port Pirie, Kosovo).

For these reasons it is hardly surprising that the results are discrepant. The results of the seven prospective studies are inconsistent if, as is necessary in the present context, the effect only of the prenatal and perinatal exposure to lead is considered as the predictor for cognitive development. Three of the studies revealed no such relationship (Glasgow, Port Pirie, Sydney). In three other studies (Boston, Cleveland, Cincinnati) a significant negative association was determined at the age of six months. In the Kosovo study this association was detected in samples from children aged 48 months but not from younger children. In four of the studies (not in Sydney, Glasgow or Cleveland) mainly relationships between cognitive development and postnatal lead exposure were found and these are in accordance in this respect with the results of most cross-sectional studies (not discussed here).

Thus it is still not clear whether specifically prenatal or perinatal exposure to lead is relevant for a persistent impairment of later cognitive development; the clearest evidence was obtained in the Boston study in which 100 µg/l blood was the effect threshold (Bellinger *et al.* 1991). The impairment persists, however, for at most 24 months. Prenatal and postnatal blood lead levels are necessarily correlated, so that a negative correlation between prenatal or perinatal blood lead levels and postnatal development could also be detected postnatally. Then it would be justified to doubt that maternal exposure to lead can affect the later cognitive development of her child and to conclude that the developmental toxicological data indicate that changes in the BAT value for women of child-bearing age are not necessary. However, the quantitative situation is unclear and more epidemiological studies are necessary" (Greim 2002).

The available literature shows that a scientifically grounded value, based on a threshold concentration for foetal effects, still cannot be evaluated. Effects can be seen even at low levels of exposure; a threshold concentration for such effects can probably not be determined. This view is further supported by the most recent studies (Canfield *et al.* 2003, Rogan and Ware 2003, Selevan *et al.* 2003, Walkowiak *et al.* 1998).

10.2 Background exposure

Exposure situation in Germany
The internal exposure to lead in the general population has decreased considerably in recent years. The following reference values were given for the German population for the years 1990/92 (95th percentiles) by the "Kommission Human-Biomonitoring des Umweltbundesamtes" (Federal Environmental Agency, UBA) (Krause *et al.* 1996):
– Women aged 25 to 69 years: 90 µg/l
– Men aged 25 to 69 years: 120 µg/l

In the meantime, the level of lead in blood has decreased further in Germany. The environmental survey from 1998 yielded the following reference values (95th percentiles):

- Women aged 18 to 69 years: 62 µg/l
- Men aged 18 to 69 years: 79 µg/l

The median value was in the range of 27 to 36 µg/l (Becker et al. 2002).

The Federal Environmental Agency commission therefore set the following reference values on the basis of the 3rd environmental survey (Becker et al. 2002, UBA 2002):
- Women (18–69 years): 70 µg/l *
- Men (18–69 years): 90 µg/l *

International situation

In the "Tracy project", in 1996 the blood lead levels in the general population worldwide were evaluated. Ten studies were selected which were suitable for evaluation as regards the analysis and definition of the population group. They concluded that no general reference values for the blood lead level can be given, as the lead concentrations in blood depend greatly on the sampling time. As a result of the elimination of lead from motor fuels, there has been a constant decrease in blood lead levels (Herber 1999). In Sweden, for example, they fell from 60 µg lead per litre blood in 1978 to 25 µg/l in a period of 15 years. The study showed that numerous factors, such as age, sex, ethnic origin, food, drinking and smoking habits, hobbies, the season and year of sampling, residential area and geographical origin influence the lead levels in the general population (Gerhardsson et al. 1996). In the NHANES (National Health and Nutrition Examination Survey) report of the CDC (2003), for the female American general population a 95th percentile of 40 µg/l blood was evaluated. This value is based on the analysis of 4057 blood samples, taken in the period 1999–2000 in the United States.

11 Re-evaluation of the BAT Value

For a re-evaluation of the BAT value for women younger than 45 years the following points are relevant:
- As for the BAT value for women older than 45 years and for men, only the lead concentration in blood can be used as a biomarker. The exposure range in discussion disqualifies the use of effect parameters, such as the excretion of δ-aminolaevulinic acid in urine (ALA-U) and the concentration of the erythrocyte porphyrins, for occupational-medical health monitoring. Determination of the activity of ALA-D (δ-aminolaevulinic acid dehydratase) in blood as a monitoring parameter is not practicable.
- A scientifically grounded BAT value for the blood lead level in women of child-bearing age cannot at present be evaluated, as a clear threshold concentration cannot be defined. In the Boston longitudinal study, however, an effect threshold of 100 µg/l is assumed (Bellinger et al. 1991). The need for a stricter limitation of exposure for women of child-bearing age (below 45 years) compared to for women not of child-bearing age (45 years and above) and men is evident for the reasons given above and

---

* When using the updated reference values, an analytical uncertainty of ± 20 % must be taken into account.

from a preventive point of view. As already stipulated in the occupational-medical, toxicological documentation for the previous BAT value of 300 µg/l blood, in view of the general considerations women of child-bearing age should not be exposed to a much greater degree than the average population (Henschler and Lehnert 1989). On the basis of the environmental survey from 1998, currently for the German general population a blood lead concentration for women (18–69 years of age) of 62 µg/l is given as the 95th population percentile. A blood lead concentration of 70 µg/l is the resulting reference value.

In analogy to the BAT value for women younger than 45 years evaluated in 1987, on the basis of the current background exposure of the general population and taking into consideration the findings in a longitudinal study which gives an effect threshold of 100 µg/l, a BAT value has been set of

100 µg lead per l blood

12 References

Baghurst PA, McMichael AJ, Wigg NR, Vimpani GV, Robertson EF, Roberts RJ, Tong SL (1992) Environmental exposure to lead and children's intelligence at the age of seven years. The Port Pirie cohort study. New Engl J Med 327: 1279–1284

Becker K, Kaus S, Krause C, Lepom P, Schulz C, Seiwert M, Seifert B (2002) German environmental survey 1998 (GerES III): environmental pollutants in blood of the German population. Int J Hyg Environ Health 205: 297–308

Bellinger DC, Needleman HL, Leviton A, Waternaux C, Rabinowitz MB, Nichols ML (1984) Early sensory-motor development and prenatal exposure to lead. Neurobehav Toxicol Teratol 6: 387–402

Bellinger D, Leviton A, Needleman H, Waternaux C, Rabinowitz M (1986) Low-level lead exposure and infant development in the first year. Neurobehav Toxicol Teratol 8: 151–161

Bellinger D, Leviton A, Waternaux C, Needleman H, Rabinowitz M (1987) Longitudinal analyses of perinatal and postnatal lead exposure and early cognitive development. N Engl J Med 316: 1037–1043

Bellinger D, Sloman J, Leviton A, Rabinowitz M, Needleman HL, Waternaux C (1991) Low-level lead exposure and children's cognitive function in the preschool years. Pediatrics 87: 219–227

Canfield RL, Henderson Jr. CR, Cory-Slechta DA, Cox C, Jusko TA, Lanphear BP (2003) Intellectual impairment in children with blood lead concentrations below 10 µg per deciliter. N Engl J Med 348: 1517–1526

CDC (Centers for Disease Control and Prevention) (2003) Second national report on human exposure to environmental chemicals. NCEH Pub. No. 02-0716, National Center for Environmental Health, Division of Laboratory Sciences, Atlanta, Georgia

Cooney GH, Bell A, Mcbride W, Carter C (1989) Neurobehavioural consequences of prenatal low level exposure to lead. Neurotoxicol Teratol 11: 95–104

Dietrich KN, Krafft KM, Bornschein RL, Hammond PB, Berger O, Succop PA, Bier M (1987) Low level fetal lead exposure effect on neurobehavioral development in early infancy. Pediatrics 80: 721–730

Dietrich KN, Succop PA, Bornschein RL, Krafft KM, Berger O, Hammond PB, Buncher CR (1990) Lead exposure and neurobehavioral development in later infancy. Environ Health Perspect 89: 13–19

Dietrich KN, Succop PA, Berger OG, Hammond PB, Bornschein RL (1991) Lead exposure and the cognitive development of urban preschool children: the Cincinnati lead study cohort at age 4 years. Neurotoxicol Teratol 13: 203–211

Dietrich KN, Succop PA, Berger OG, Keith RW (1992) Lead exposure and the central auditory processing abilities and cognitive development of urban children: the Cincinnati lead study cohort at age 5 years. Neurotoxicol Teratol 14: 51–56

Dietrich KN, Berger OG, Succop PA, Hammond PB, Bornschein RL (1993) The developmental consequences of low to moderate prenatal and postnatal lead exposure: intellectual attainment in the Cincinnati lead study cohort following school entry. Neurotoxicol Teratol 15: 37–44

Ernhart CB, Wolf AW, Kennard MJ, Erhard P, Filipovich HF, Sokol RJ (1986) Intrauterine exposure to low levels of lead: the status of the neonate. Arch Environ Health 41: 287–291

Ernhart CB, Morrow-Tlucak M, Marler MR, Wolf AW (1987) Low level lead exposure in the prenatal and early preschool periods: early preschool development. Neurotoxicol Teratol 9: 259–270

Ernhart CB, Morrow-Tlucak M, Wolf AW (1988) Low level lead exposure and intelligence in the preschool years. Sci Total Environ 71: 453–459

Ernhart CB, Morrow-Tlucak M, Wolf AW, Super D, Drotar D (1989) Low level lead exposure in the prenatal and early preschool periods: intelligence prior to school entry. Neurotoxicol Teratol 11: 161–170

Gerhardsson L, Kazantzis G, Schutz A (1996) Evaluation of selected publications on reference values for lead in blood. Scand J Work Environ Health 22: 325–331

Greim H (Ed.) (2002) Lead and its inorganic compounds, apart from lead arsenate and lead chromate. in: Occupational toxicants, Vol. 17, Wiley-VCH, Weinheim

Henschler D, Lehnert G (Eds) (1989) Arbeitsmedizinisch-toxikologische Begründungen von BAT-Werten, 4. Lieferung, VCH, Weinheim

Herber RFM (1999) Review of trace element concentrations in biological specimens according to the TRACY protocol. Int Arch Occup Environ Health 72: 279–283

Krause C, Babisch W, Becker K, Bernigau W, Helm D, Hoffmann K, Nöllke P, Schulz C, Schwabe R, Seiwert M, Thefeld W (1996) Umwelt-Survey 1990/92, Band Ia: Studienbeschreibung und Human-Biomonitoring: Deskription der Spurenelementgehalte in Blut und Urin der Bevölkerung der Bundesrepublik Deutschland. WaBoLu-Hefte 1/1996, Werbung und Vertrieb, Berlin

McMichael AJ, Baghurst PA, Wigg NR, Vimpani GV, Robertson EF, Roberts RJ (1988) Port Pirie cohort study: environmental exposure to lead and children's abilities at the age of four years. N Engl J Med 319: 468–475

Moore MR, Bushnell IWR, Goldberg Sir A (1989) A prospective study of the results of changes in environmental lead exposure in children in Glasgow. in: Smith MA, Grant LD, Sors AI (Eds) Lead exposure and child development. An international assessment. Published for the Commission of the European Community and the US Environmental Protection Agency, Kluwer Academic Publishers, Dordrecht, 371–378

Pocock SJ, Smith M, Baghurst P (1994) Environmental lead and children's intelligence: a systematic review of the epidemiological evidence. BMJ 309: 1189–1197

Rogan WJ, Ware JH (2003) Exposure to lead in children – how low is low enough? N Engl J Med 348: 1515–1516

Selevan SG, Rice DC, Hogan KA, Euling SY, Pfahles-Hutchens A, Bethel J (2003) Blood lead concentration and delayed puberty in girls. N Engl J Med 348: 1527–1536

UBA (Umweltbundesamt) (2002) Bekanntmachungen des Umweltbundesamtes – Addendum zur "Stoffmonographie Blei – Referenz- und Human-Biomonitoring-Werte" der Kommission "Human-Biomonitoring" Stellungnahme der Kommission "Human-Biomonitoring" des Umweltbundesamtes. Bundesgesundheitsbl Gesundheitsforsch Gesundheitsschutz 45: 752–753

Vimpani GV, Baghurst PA, Wigg NR, Robertson EF, McMichael AJ, Roberts RR (1989) The Port Pirie cohort study – cumulative lead exposure and neurodevelopmental status at age two years: Do HOME scores and maternal IQ reduce apparent effects of lead on Bayley mental scores? in: Smith MA, Grant LD, Sors AI (Eds) Lead exposure and child development. An international assessment. Published for the Commission of the European Community and the US Environmental Protection Agency, Kluwer Academic Publishers, Dordrecht, 332–344

Walkowiak J, Altmann L, Krämer U, Sveinsson K, Turfeld M, Weishoff-Houben M, Winneke G (1998) Cognitive and sensorimotor functions in 6-year-old children in relation to lead and mercury levels: adjustment for intelligence and contrast sensitivity in computerized testing. Neurotox Teratol 20: 511–521

Wasserman G, Graziano JH, Factor-Litvak P, Popovac D, Morina N, Musabegovic A, Vrenezi N, Capuni-Paracka S, Lekic V, Preteni-Redjepi E, Hadzialjevic S, Slavkovich V, Kline J, Shrout P, Stein Z (1992) Independent effects of lead exposure and iron deficiency anemia on developmental outcome at age 2 years. J Pediatr 121: 695–703

Wasserman GA, Graziano JH, Factor-Litvak P, Popovac D, Morina N, Musabegovic A, Vrenezi N, Capuni-Paracka S, Lekic V, Preteni-Redjepi E, Hadzialjevic S, Slavkovich V, Kline J, Shrout P, Stein Z (1994) Consequences of lead exposure and iron supplementation on childhood development at age 4 years. Neurotoxicol Teratol 16: 233–240

WHO (World Health Organization) (1995) Inorganic lead. Environmental Health Criteria 165, Geneva

Wigg NR, Vimpani GV, McMichael AJ, Baghurst PA, Robertson EF, Roberts RJ (1988) Port Pirie cohort study: childhood blood lead and neuropsychological development at age two years. J Epidemiol Community Health 42: 213–219

Authors: K.H. Schaller, H.M. Bolt
Approved by the Working Group: 23.07.2003

Table 1. Prospective studies of toxic effects on reproduction in women (Greim 2002)

Authors, year of publication	Country	Test persons Age (years)	N	Lead exposure	Variables	Results
Bellinger et al. 1984, 1986, 1987, 1991	Boston, MA	birth to 10	249	level of lead in umbilical cord blood (low: 1–8 µg/dl, high: 14.6 µg/dl) 6, 12, 24 months (low: < 10 µg/dl, high: > 10 µg/dl)	Bayley scales (MDI, PDI), McCarthy scales GCI, WISC-R	1) significant inverse relationship between the level of lead in umbilical cord blood and the Bayley MDI at all times up to 24 months; 2) non-significant relationships between the blood lead level and the Bayley MDI at the age of 6 months; 3) significant negative relationship between the 24 month blood lead level and the 57 month McCarthy scale; the prenatal effects of lead remain significant at 57 months for low SES children, for those with a simultaneous higher blood lead level and for boys; 4) inverse relationship between the WISC-R and the blood lead level at 24 months
Dietrich et al. 1987, 1990, 1991, 1992, 1993	Cincinnati, OH	birth to 6.5	305	level of lead in umbilical cord blood (x = 8 µg/dl), blood lead level (24–60 months)	Bayley scales, K-ABC, WISC-R	significant negative relationship between the umbilical cord lead level and 6 month Bayley MDI, not significant at the age of 24 months, K-ABC not significant at 4 years, significant at 5 years; inverse relationship at the age of 6.5 years between the WISC-R and the postnatal blood lead level
Ernhart et al. 1986, 1987, 1988, 1989	Cleveland, OH	birth, follow-up studies at 6 months, 1, 2, 3 and 4 years and 10 months	162	level of lead in umbilical cord blood (x = 5.8 µg/dl), blood lead level (6 months to 4 years)	Brazelton Neonatal Behavior Assessment Scale, Graham/ Rosenblith test for newborn babies, Bayley scales, WPSSI	prenatal lead level negatively significant for 2 neonatal parameters (others not significant); with 6 month Bayley scale, not significant with 1, 2, 3 year Bayley scale and the WPPSI after adjustment for confounders
Moore et al. 1989	Glasgow, UK	birth, maternal PbB, follow-up study (24 months)	151	high: > 300 µg/l medium: 150–200 µg/l low: < 100 µg/l	Bayley scales	no significant group differences at the age of 12 and 4 months

Table 1. continued

Authors, year of publication	Country	Test persons Age (years)	N	Lead exposure	Variables	Results
Wigg et al. 1988 McMichael et al. 1988, Baghurst et al. 1992, Vimpani et al. 1989	Port Pirie, Australia	birth to 7	537	level of lead in umbilical cord blood ($x = 8.8$ µg/dl), blood lead level over 7 years, level of lead in teeth at the age of 7 years	Bayley scales, McCarthy GCI, WISC-R	significant inverse relationship between the 6 month blood lead level and 2 year Bayley MDI, relationship between the level of lead in umbilical cord blood and the Bayley MDI not significant; significant inverse relationship between the blood lead at the age of 2 and 3 years and McCarthy GCI at the age of 4 years; the index for the cumulative blood lead concentration shows the greatest inverse relationship to the GCI score at the age of 4 years; the WISC-R has a significant inverse relationship to the level of lead in blood and teeth at the age of 7 years
Wasserman et al. 1992, 1994	Kosovo (formerly Yugoslavia), towns with smelting works and towns with low lead exposure	birth to 4	392	lead level ($x = 35.5$ µg/dl smelting towns) ($x = 8.4$ µg/dl towns with low lead exposure)	Bayley MDI, McCarthy GCI	lead effects (confounded) with anaemia, lead effects lower than the anaemia effects at the age of 2 years; significant lead–GCI relationship at the age of 4 years; lead-related behavioural disturbances
Cooney et al. 1989	Sydney, Australia	birth to 5	318	level of lead in umbilical cord blood ($x = 8.1$ µg/dl)	Bayley MDI	non-significant relationship between blood lead level and MDI after covariable adjustment

Abbreviations: BNBAS: Brazelton Neonatal Behaviour Assessment Scale; BSID: Bayley Scales of Infant Development; GCI: General Cognitive Index; K-ABC: Kaufman-Assessment Battery of Children; MDI: Mental Development Index; MSCA: McCarthy Scales of Children's Abilities; PDI: Psychomotor Development Index; SES: Socio Economic Status; WISC-R: Wechsler Intelligence Scale for Children-Revised; WPPSI: Wechsler Preschool and Primary Scale of Intelligence; PbB: blood lead level

Lead and its compounds (except lead arsenate, lead chromate and alkylated compounds)

BLW for women older than 45 years for men	400 µg lead/l blood Sampling time: no restrictions
Date of evaluation	2005
BLW for women younger than 45 years	100 µg lead/l blood Sampling time: no restrictions
Date of evaluation	2005

13 Evaluation of BLW Values

Lead has been classified in Category 3 B for carcinogenic substances in the workplace due to occurrence of genotoxic effects. In addition to genotoxic effects there is also a suspicion of occurrences of tumours, which is discussed in some of the studies. The previous MAK value of 100 µg/m^3 has been withdrawn, and thus the BAT value has to be withdrawn, too.
In the case of lead however, one must consider that the previous BAT values of 400 µg/l blood for women older than 45 years and for men, and the BAT value of 100 µg/l blood for women younger than 45 years are based on toxic effects. This includes neurotoxic (and also behavioural) effects, which are of considerable relevance for the occupational hygiene concerns. The present BAT values have therefore been carried over as BLW values. Especially the prevention of neurotoxic effects is of great importance for lead-exposed persons and can be achieved by observing the BLW values. This also applies to the minimizing of reproductive toxic effects. Therefore, the BLW values have been set at

400 µg lead/l blood (for women older than 45 years and for men)
100 µg lead/l blood (for women younger than 45 years)

The importance of occupational-medical health surveillance including biological monitoring has to be stressed.

Author: H.M. Bolt
Approved by the Working Group: 23.05.2005

The MAK-Collection Part II: BAT Value Documentations, Vol. 4. DFG, Deutsche Forschungsgemeinschaft
Copyright © 2005 WILEY-VCH Verlag GmbH & Co. KGaA, Weinheim
ISBN: 3-527-27049-3

Manganese and its inorganic compounds

BAT	20 µg manganese/l blood Sampling time: end of exposure or end of shift, with long-term exposure after several previous shifts
Date of evaluation	2001
CAS No.	7439-96-5
Formula	Mn
Molecular weight	54.94
Melting point	1244°C
Boiling point	2032°C
Density at 20°C	7.20–7.40 g/cm^3
MAK [last established: 1994]	0.5 mg/m^3 I (measured as the inhalable fraction of the aerosol)

Until 2001 there was no BAT value for manganese and its inorganic compounds. The MAK value was lowered in 1994 from 5 mg/m^3 to 0.5 mg/m^3. There is detailed documentation available (Greim 1999). The results showed that effects on the central nervous system after inhalation of manganese dust are the critical toxic effects in man. Numerous studies from the period 1980 to 1994 showed that slight neurological symptoms are observed below 5 mg/m^3. Although a LOAEL (lowest observed adverse effect level) of about 0.25 mg/m^3 was found in a Swedish study, after considering the detection characteristics of the dust sampling devices and the fact that the symptoms did not develop in all exposed persons and were not dose-dependent, the MAK value was set at 0.5 mg/m^3 (Greim 1999).

1 Metabolism and Kinetics

There are numerous studies and reviews available of the toxicokinetics of manganese and inorganic manganese compounds which serve as the basis for the following documentation (Alessio and Lucchini 1997, Barceloux 1999, Greim 1999, Zidek 1983, Zimmermann-Hölz 1989).

1.1 Absorption

Under workplace conditions manganese is absorbed mainly via inhalation and to a smaller extent via the gastrointestinal tract. Dermal absorption of metallic manganese and inorganic manganese compounds is not known. The solubility of the manganese compounds is decisive for absorption. Manganese oxides are less soluble than manganese salts. Absorption in the lungs is also affected by the particle size. Small particles can reach the alveoli, large particles are absorbed in the upper respiratory passages and can be eliminated by the mucociliary transport mechanism and then swallowed. Around 40 % of the manganese particles that reach the alveoli are absorbed (Oberdoerster and Cherian 1988).

1.2 Distribution and accumulation

The amount of manganese in the body of a 70 kg adult is estimated to be 10 to 20 mg. The concentration in the tissues varies considerably and makes up between 0.1 and 1 µg/g wet weight; the concentrations are somewhat higher in the liver, pancreas, kidneys and brain (WHO 1981).

The highest manganese concentrations were measured in the basal ganglia (*nucleus caudatus*, *globus pallidus* and putamen) (Larsen et al. 1979, Smeyers-Verbeke et al. 1976). After intratracheal administration, similarly high concentrations were found in the various tissues for manganese chloride and manganese oxide, despite their different solubility.

Manganese is bound to erythrocytes in the form of a manganese porphyrin and in plasma to proteins (transferrin, macroglobulin). It can pass the placenta and the blood-brain barrier.

The bivalent and trivalent oxidation states (Mn^{2+} and Mn^{3+}) and the reaction of Mn^{3+} with transferrin are important for the toxicokinetic properties (Aschner et al. 1999).

1.3 Elimination

Manganese is eliminated mainly with the faeces (about 90 %) and in small amounts with the urine (about 6 %). The amount eliminated with the faeces is about 2 to 3 mg. After inhalation of manganese chloride or manganese oxide (Mn_3O_4), around 60 % of the manganese deposited in the lungs is eliminated after 4 days (Mena et al. 1969).

Araki et al. (1986) reported that manganese mainly undergoes glomerular filtration; tubular secretion or reabsorption do not take place.

1.4 Biological half-times

The half-times mainly depend on the routes of absorption. In animal experiments, after inhalation, half-times of 223 and 267 days were determined in the brain (Newland *et al.* 1987).

In studies with radioactively-labelled manganese chloride administered intravenously to test persons, it was determined with whole body X-rays that elimination is a function of long-term manganese exposure (Cotzias *et al.* 1968, Mena *et al.* 1967).

Bader *et al.* (1999) investigated ten persons formerly occupationally exposed to manganese dioxide in dust and no longer exposed for at least 6 months, and seven persons not exposed. No differences were found between the two groups for the manganese concentrations in blood and urine.

2 Critical Toxicity

Manganese is an essential trace element. It is a cofactor for the activity of various enzymes (e.g. pyruvate carboxylase, arginase, phosphatase, superoxide dismutase, glutamine synthetase, manganese-dependent ATPase). For manganese and inorganic manganese compounds, allergic, carcinogenic and genotoxic effects have not been consistently detected to date. The studies with animals of the toxic effects on reproduction were interpreted as yielding evidence that prenatal toxicity is not to be expected with observance of the MAK value (Greim 1999).

The critical toxicity of manganese and its inorganic compounds is determined by the effects on the respiratory passages and lungs, and the central nervous system.

The inhalation of manganese, e.g. in the form of manganese dioxide, can, in high concentrations, lead to acute respiratory symptoms and even to symptoms similar to pneumonia. These are comprehensively documented in early occupational-medical publications (Baader 1954, Koelsch 1963).

The occurrence of pneumonia has been reported in connection with exposure to high levels of manganese dust of up to 800 mg/m^3 (Rodier 1955). In a Belgian study, in addition to investigating neuropsychological and biological parameters, also lung function analyses were carried out (Roels *et al.* 1992). The authors concluded that slight respiratory symptoms can develop even at manganese concentrations as low as 1 mg/m^3.

With regard to the neurotoxicity of manganese, it must first be noted that for the peripheral nervous system (e.g. polyneuropathy) and/or the autonomous nervous system manganese-specific effects have not been consistently described.

Since they were first described by Couper (1837), there have been numerous reports of Parkinson-like symptoms after long-term exposure to manganese. These are summarized in various reviews (Alessio and Lucchini 1997, Chu *et al.* 1995, Cook *et al.* 1974, Pal *et al.* 1999, WHO 1981).

The disease is described as "manganism" and according to the WHO divided into three phases (WHO 1987):

I. Symptoms such as headaches, memory disorders, dizziness, weakness and fatigue dominate, psychotic episodes can occur. This phase can last for several months.
II. Neurological symptoms develop. These include unsteady gait, stuttering, compulsive crying and laughing, muscle dystonia (increased resting tonus). The symptoms usually last for several months.
III. The neurological deficits persist even after the end of exposure.

The similarity between manganism and Parkinson's disease seems at first to be limited to the symptoms, but is further substantiated by the detection of lesions in the basal ganglia, in particular in the *globus pallidus* (McMillan 1999). Today it is known that there are clear pathophysiological differences between manganism and Parkinson's disease. Parkinson's disease, for example, is the result of the degeneration of the nigrostriatum. This finding is rarely detected in patients with manganism (Wolters *et al.* 1989). Another characteristic is the fact that Parkinson's disease does not respond to L-dopa (Pal *et al.* 1999).

The neurotoxic mechanism is still largely unclear (Chu *et al.* 1995).

The finding that the extrapyramidal regions of the CNS have a special pathogenetic role is supported by more recent studies with imaging procedures (Chu *et al.* 1995, Dietz and Schunk 1998, Dietz *et al.* 2000, Pal *et al.* 1999, Triebig 1998).

There are numerous publications and more recent reviews on the question of the chronic neurotoxicity of manganese (Chu *et al.* 1995, Dietz and Schunk 1998, Iregren 1994, 1999, Mergler and Baldwin 1997). Important conclusions are:

1. Motor disturbances and reduced psychomotor performance are sensitive and relatively specific. These include the following functions: hand tremor, rapid finger movements, and simple hand movements, which can be reliably measured using standardized procedures.
2. The relevant psychological functions include memory performance, reaction time and complex cognitive performance.
3. Subjective symptoms and impaired well-being are not specific enough to be able to draw concrete conclusions from them.

For long-term effects, the following studies are available. In former mine workers in Chile, who were investigated after an average exposure period of 20 years and were no longer exposed for at least 5 years, tremor and slow movement were diagnosed (Hochberg *et al.* 1996). The authors conclude that it is unclear whether the persons already had these symptoms earlier or whether they had developed recently.

Roels *et al.* (1999) reported that after a period without exposure of at least 3 years, the co-ordination disturbances determined earlier could no longer be detected. For the variables tremor (steadiness) and simple reaction time, however, no improvement was detected (Roels *et al.* 1999).

Crump and Rousseau (1999) reported about employees from manganese oxide and salt production, who underwent regular neurological examination in the period from 1985 to 1996. There was no evidence that the neurological symptoms and psychological effects initially determined increased, despite continued exposure, after age and the time of investigation were taken into account.

3 Relationship between Internal Exposure and Effects

Summarized below are those studies which describe the neurotoxic effects of exposure to manganese. To be able to evaluate a BAT value, a meta-analysis was carried out; the main conclusions are given below.

27 studies were identified in which psychomotor or neuropsychological test procedures were used.

Only seven studies fulfilled the requirements of the meta-analysis with regard to the manganese concentrations in blood (Lucchini *et al.* 1995, 1999, Maroni 1997, Mergler *et al.* 1994, Roels *et al.* 1987, 1992, Sjögren *et al.* 1996).

A meta-analysis of the studies with manganese in urine was not possible as no standard deviations were given (Lucchini *et al.* 1997, 1999, Maroni 1997, Roels *et al.* 1987, 1992).

Although neuropsychological or psychomotor methods were used, 13 studies could not be considered, either because the methods or the exposure ranges were not comparable (Amr *et al.* 1993, Beuter *et al.* 1994, Chandra *et al.* 1981, Chia *et al.* 1993, Emara *et al.* 1971, Ferraz *et al.* 1988, Hochberg *et al.* 1996, Hua and Huang 1991, Kaji *et al.* 1993, Kondakis *et al.* 1989, Lucchini *et al.* 1997, Mergler *et al.* 1998, Seeber *et al.* 1984).

The results of studies in which the manganese exposure of the group was in the range in question but, as a result of the study design, the description of the results or the exposure range of the control group could not be evaluated by meta-analysis, were taken into consideration separately (Crump and Rousseau 1999, Dietz *et al.* 2003, Gibbs *et al.* 1999, Iregren *et al.* 1990, Kaji *et al.* 1993, Roels *et al.* 1999, Siegl and Bergert 1982, Wennberg *et al.* 1991, 1992).

Table 1 shows the mean levels of manganese exposure found in the studies included in the meta-analysis.

Table 1. Studies included in the meta-analysis

	Persons exposed to manganese				Control group	
		Duration of exposure (years)	Manganese in blood ($\mu g/l$)			Manganese in blood ($\mu g/l$)
	N	Mean	Mean	Maximum	N	Mean
Lucchini *et al.* 1995	20	13.8	11.9	18	19	6.0
Lucchini *et al.* 1999	61	15.2	9.7	19	87	6.0
Maroni 1997	154	10.8	10	74	104	7
Mergler *et al.* 1994	74	19.4	10.3	13*	74	6.8
Roels *et al.* 1987	141	7.1	12.2	36	104	4.9
Roels *et al.* 1992	92	5.3	8.1	21	101	6.8
Sjögren *et al.* 1996	12	19.5	8.5	14	39	7

* upper quartile

The internal exposure to manganese covered a large range; maximum values up to 74 µg/l were determined.

Table 2 lists those psychomotor test procedures which were used in at least three studies.

Table 2. Neuropsychological and motor test procedures used in the studies

	Lucchini et al. 1995	Lucchini et al. 1999	Maroni 1997	Mergler et al. 1994	Roels et al. 1987	Roels et al. 1992	Sjögren et al. 1996
Psychomotor test procedures:							
reaction time	+	+	+	+	+	+	+
tapping	+	+	–	+	–	–	+
steadiness (tremor)	–	–	–	+	+	+	–
Neuropsychological procedures:							
digit-symbol test	+	+	+	+	–	–	+
digit span	+	+	–	+	–	–	+

The test procedures listed in Table 2 were taken into consideration in the meta-analyses. In addition, the results of complaint questionnaires and test procedures for simple motor hand movements were evaluated.

3.1 Reaction time

Table 3 shows the mean values and standard deviations of the simple reaction times in the various studies.

Table 3. Mean values and standard deviations of the simple reaction times

Authors	Exposed persons			Control group		Results of the significance test
	Manganese in blood (µg/l)	N	Simple reaction time	N	Simple reaction time	
Lucchini et al. 1995	11.9	20	242 ± 24	19	237 ± 16	not sign.
Maroni 1997	10	154	241*	104	245*	not sign.
Mergler et al. 1994	10.3	74	270 ± 34	74	269 ± 28	not sign.
Roels et al. 1987	12.2	141	272 ± 31	104	263 ± 29	significant
Roels et al. 1992	8.1	92	265 ± 28	101	249 ± 20	significant
Sjögren et al. 1996	8.5	12	242 ± 20	39	229 ± 23	not sign.

* no standard deviation given by the author

The simple reaction time was found to be significantly dependent on the manganese concentration in blood in only few studies. The highest effect sizes were found, however, with low manganese concentrations in blood. An exposure–effect relationship could not be detected.

3.2 Tapping

For tapping (rapid tapping with a finger or peg on a board or button) most studies differentiated between tapping with the dominant hand and the non-dominant hand. The results are shown in Table 4.

Table 4. Mean values and standard deviations for tapping. Given is the mean number of "hits"

Authors	Exposed persons			Control group		Results of the significance test
	Manganese in blood (µg/l)	N	Tapping	N	Tapping	
Lucchini et al. 1995	11.9	20	64.7 ± 5.5	19	69.1 ± 8.1	significant
Mergler et al. 1994						
dominant hand	10.3	74	56.4 ± 7.8	74	54.2 ± 7.9	not sign.
non-dominant hand	10.3	74	51.6 ± 7.8	74	49.6 ± 7.1	not sign.
Sjögren et al. 1996						
dominant hand	8.5	12	63.7 ± 7.1	39	66.1 ± 8.3	not sign.
non-dominant hand	8.5	12	54.3 ± 7.7	39	62.3 ± 9.3	significant

Effects were found in only few studies. An exposure–effect relationship could not be detected.

3.3 Tremor (steadiness)

Hand tremor was tested using various devices. The test persons had to insert a peg into small holes with different radii and keep it as still as possible. The results are device-specific and refer to the number of times the peg touched the wall or the duration of the contact. They are shown in Table 5.

Table 5. Mean values and standard deviations for the number of errors or the duration of the error in tremor analyses

Authors	Exposed persons			Control group		Results of the significance test
	Manganese in blood (µg/l)	N	Tremor	N	Tremor	
Lucchini et al. 1999	9.7	61	7.3 ± 1.2	83	7.0 ± 0.7	significant
Mergler et al. 1994	10.3	74	60 ± 37	74	47 ± 41	significant
Roels et al. 1992	8.1	92	16 ± 9	101	13 ± 9	significant

In some other studies, not included in the meta-analysis (Roels et al. 1987, Crump and Rousseau 1999), significant relationships between manganese exposure and tremor were reported. However, no exposure–effect relationships were found between these studies.

3.4 Simple hand movements

In some studies simple hand movements were tested. Although meta-analysis cannot be carried out, as a result of the different procedures used, the results of the diadochokinesis tests and the pegboard test are reported below.

Beuter et al. (1994) describe functional asymmetry between the right and left hand in employees exposed to manganese. In Wennberg et al. (1991, 1992) the test group were found to have a lower frequency with the dominant and non-dominant hand than the control group. Sjögren et al. (1996), however, did not observe any significant results.

Thus, significant effects were found in two of three studies of diadochokinesis. These refer, however, to different hand functions.

Pegboard tests measure manual dexterity. The task is to insert small pegs into holes on a board as quickly as possible.

Table 6 shows the results of the pegboard tests. Given are the number of pegs achieved.

Table 6. Mean values and standard deviations for the pegboard tests

Authors	Exposed persons			Control group		Results of the significance test
	Manganese in blood (µg/l)	N	Tremor	N	Tremor	
Dietz et al. 2003						
dominant hand	14.6	11	42.6 ± 5.0	11	42.7 ± 4.2	not sign.
non-dominant hand	14.6	11	43.7 ± 5.8	11	43.8 ± 4.1	not sign.
Mergler et al. 1994						
dominant hand	10.3	74	13.7 ± 1.9	74	13.7 ± 1.5	not sign.
non-dominant hand	10.3	74	13.4 ± 1.8	74	13.3 ± 1.6	not sign.
Sjögren et al. 1996						
dominant hand	8.5	12	42.5 ± 2.7	39	45.3 ± 3.7	significant
non-dominant hand	8.5	12	39.0 ± 2.5	39	41.8 ± 3.7	significant
Wennberg et al. 1991	–	30	26.0*	60	26.8*	not sign.

* no standard deviation given by the authors

In most studies no significant effects were found in the pegboard test. Only Sjögren et al. (1996) describe, despite relatively low manganese concentrations in the blood, a significant difference between exposed persons and the control group.

3.5 Digit-symbol test

In the digit-symbol test, digits are assigned to different symbols. It tests psychomotor speed and is regarded as a measure for concentration (Tewes 1991).

In Table 7 different values for the digit-symbol test are given, as the authors do not describe the results in the same way. In the studies of Maroni (1997) and Mergler et al. (1994) the number of correct items are given, in the studies of Lucchini et al. (1995) and Sjögren et al. (1996) the latency periods.

Table 7. Mean values and standard deviations for the digit-symbol test

Authors	Exposed persons			Control group		Results of the significance test
	Manganese in blood (µg/l)	N	Digit-symbol test	N	Digit-symbol test	
Lucchini et al. 1995	11.9	20	4.0 ± 0.7	19	3.2 ± 0.7	significant
Maroni 1997	10	154	29*	104	27*	not sign.
Mergler et al. 1994	10.3	74	47.0 ± 8.0	74	47.6 ± 9.3	not sign.
Sjögren et al. 1996	8.5	12	3.2 ± 0.4	39	2.9 ± 0.6	not sign.

* no standard deviation given by the author

Significant effects between the manganese concentrations in blood and the digit-symbol test were reported only in few cases. The significant effects occur in association with a comparatively high manganese concentration or long-term exposure.

3.6 Digit span

The test "repeating digits" measures short-term memory. It is sensitive for persons with organic diseases of the brain (Lehrl et al. 1992). In the different versions of the test procedure, either numbers or letters must be reproduced which were presented shortly before.

Table 8 shows the results of the digit span test in the various studies.

Table 8. Mean values and standard deviations for the digit span test

Authors	Exposed persons			Control group		Results of the significance test
	Manganese in blood (µg/l)	N	Digit span	N	Digit span	
Lucchini et al. 1995	11.9	20	6.7 ± 1.2	19	7.9 ± 1.1	significant
Lucchini et al. 1999	9.7	61	6.9 ± 1.9	83	7.1 ± 1.2	significant
Mergler et al. 1994	10.3	74	7.9 ± 1.1	74	8.1 ± 1.3	not sign.
Sjögren et al. 1996	8.5	12	6 ± 0.7	39	6.6 ± 1.4	not sign.

Statistically significant results were found only in two of the studies considered. The results therefore demonstrate there is a negative relationship between the manganese exposure and the repeating of digits as an indicator for short-term memory.

3.7 Disorders of well-being

Disorders of well-being are asked about in most studies or documented in various complaint questionnaires. Most frequently mentioned were tiredness or sleeping problems (Dietz *et al.* 2003, Mergler *et al.* 1994, Roels *et al.* 1987, Sjögren *et al.* 1996, Wennberg *et al.* 1991). In three studies irritation or aggressiveness were described (Lucchini *et al.* 1999, Mergler *et al.* 1994, Roels *et al.* 1987). In two studies the persons exposed to manganese complained more frequently of forgetfulness (Dietz *et al.* 2003, Mergler *et al.* 1994) and tinnitus (Mergler *et al.* 1994, Roels *et al.* 1987).

In the studies of Maroni (1997) and Roels *et al.* (1992) significant differences between the exposed persons and the control group were not found for any complaint. In the other studies several significant differences were found.

A uniform picture of complaints cannot be drawn up due to the dissimilarity of the complaints.

3.8 Summary of the psychomotor and neuropsychological effects

The available studies indicate that persons exposed to higher levels of manganese suffer from increased tremor and reduced performance in short-term memory.

There is also evidence that other psychomotor functions, such as reaction time and tapping, are affected. Complaints such as sleep disorders and increased irritability are mentioned more frequently (see also Iregren 1999, Mergler and Baldwin 1997).

When the possible exposure-effect relationships are considered as a whole, in particular the study of Sjögren *et al.* (1996) is conspicuous. The authors report a high effect size at relatively low concentrations of manganese in blood. As in this study, however, only 12 exposed persons were investigated, this result may be due to individual outliers. This result is therefore not included in the evaluation of possible exposure-effect relationships.

In the tests reaction time, tapping, digit-symbol test and digit span there are consistent deficits in performance in persons exposed to manganese, without there being a clear exposure-effect relationship.

For these variables the effect size increases even at average manganese concentrations as low as 10 to 12 µg/l. The only exception are the reaction times determined in the study of Roels et al. (1992). The authors found a comparably high effect size at relatively low concentrations of manganese in blood. This cannot be explained from a methodological point of view.

Overall, the effects of manganese exposure on the neuropsychological and psychomotor test procedures are, with statistical values of 0.2 to 0.4, regarded as slight (Fricke and Treinies 1985).

A NOAEL (no observed adverse effect level) and LOAEL (lowest observed adverse effect level) cannot be evaluated on the basis of the results available.

4 Relationship between External and Internal Exposure

4.1 Selection of the studies and description of the groups

The prerequisite for including published study results in the evaluation was that the number of test persons was given and also the mean or median values for external and internal exposure. If possible, also concentration ranges should be given. Of a total 23 publications, eleven had to be excluded from further evaluation as a result of the following shortcomings and incomplete data:
- the manganese was determined in serum or plasma (Chandra et al. 1981, Yiin et al. 1996),
- the data for the results of air analyses and biological monitoring were incomplete or unclear (Amr et al. 1993, Emara et al. 1971, Hua and Huang 1991, Kaji et al. 1993, Sjögren et al. 1996, Smyth et al. 1973),
- the results of air analyses and biological monitoring could only be inexactly evaluated from graphic representations, or correction factors were used (e.g. respiratory minute volume) and further calculations were necessary (Chia et al. 1993, Smargiassi et al. 1995),
- the investigations were carried out in persons no longer exposed for some time (Beuter et al. 1994).

The presentation of the results in the remaining twelve publications varied with regard to the statistical data selected. In seven publications geometric mean values were given (Buchet et al. 1993, Järvisalo et al. 1992, Lucchini et al. 1995, 1997, Mergler et al.

1994, Roels *et al.* 1987, 1992) and in four publications arithmetic mean values (Bader *et al.* 1999, Gan *et al.* 1988, Lucchini *et al.* 1999, Maroni 1997), while in one study (Zschiesche *et al.* 1986) median values were given. These different modes of representation could not be taken into account in the comparison of the studies.

In five publications the manganese concentrations in urine were related to the creatinine levels in the samples (Buchet *et al.* 1993, Järvisalo *et al.* 1992, Roels *et al.* 1987, 1992). In the correlation analyses these mean values were multiplied by the factor 1.5 to allow comparison with the volume-related results (Geigy 1985).

The number of employees exposed to manganese in the twelve studies evaluated varied between $N = 8$ (Zschiesche *et al.* 1986) and $N = 141$ (Roels *et al.* 1987). In ten studies, for comparison, between $N = 17$ (Bader *et al.* 1999) and $N = 154$ (Järvisalo *et al.* 1992) persons not occupationally exposed to manganese were investigated. In most cases the reference group was matched with the exposed group as regards age, sex, geographical origin and socio-economic status. The results of air and biological monitoring of the twelve studies are shown in Table 9.

4.2 External exposure to manganese

The level of manganese in the "inhalable dust fraction" (formerly "total dust") was used as the measure of external exposure in all studies (see Section V "Aerosols", DFG 2003). An exception are the studies of Roels *et al.* (1987) and Roels *et al.* (1992). In these cases dust sampling was carried out using a non-standardized and non-validated procedure (Pflaumbaum *et al.* 1990). A direct comparison with other methods (NIOSH, BIA) is therefore not possible (Roels *et al.* 1999).

Further data can be found in Greim (1999).

The mean concentrations of manganese in the air were usually between 0.1 mg/m^3 and 0.3 mg/m^3 (range: 0.04 mg/m^3 to 4.2 mg/m^3). In individual cases, however, also manganese concentrations up to 10 mg/m^3 (Roels *et al.* 1987, 1992) and 17 mg/m^3 (Buchet *et al.* 1993) were determined.

4.3 Correlation analyses

Using correlation analyses according to Pearson, it was tested whether there is a statistical relationship between the mean external manganese exposure and the resulting internal exposure to manganese. For this purpose, the results of the air analyses and biological monitoring from eight studies were used, in which the external exposure was at the most twice the current MAK value, i.e. 1 mg/m^3 (Bader *et al.* 1999, Järvisalo *et al.* 1992, Lucchini *et al.* 1995, 1999, Maroni 1997, Mergler *et al.* 1994, Zschiesche *et al.* 1986). To avoid distortion, the study of Gan *et al.* (1988) was excluded as a result of the high mean manganese levels in the air of over 4 mg/m^3. In addition, the studies of Roels *et al.* from 1987 and 1992, and of Buchet *et al.* (1993) were not included because the results for the inhalable dust fraction at the workplace could not be compared with those from other studies as a result of the already mentioned methodological reasons (see introductory remarks).

Figure 1 shows the mean values for the manganese concentrations in the workplace air and in the blood of the employees. In some studies several sub-groups of the exposed group were investigated. For this reason the number of points on the graph (N = 12) is greater than the number of studies (N = 7). The calculated correlation is statistically significant (r = 0.726, p < 0.001). Linear regression analysis showed that the manganese concentration in blood increases by about 1.5 µg/l per 100 µg/m^3 in the air. The point of intersection of the regression curves with the ordinate is at 8.4 µg/l and reflects the calculated manganese concentration in blood without external exposure. It correlates well with the previously deduced background exposure of about 7 to 8 µg/l blood.

Table 9. Studies of external and internal exposure to manganese

Groups	Mn in air ($\mu g/m^3$)	Mn in blood ($\mu g/l$)	Mn in urine ($\mu g/l$)[c]	Δ Mn-B (exposed/ controls)	Δ Mn-U (exposed/ controls)	Correlation Mn-A to Mn-B	Correlation Mn-A to Mn-U	Remarks	Authors
8 welders	258[m] (24–8711)	12.1[m] (7.4–33.5)	0.7[m] (0.2–6.7)	–	–	no	no		Zschiesche et al. 1986
141 smelters 104 controls	940[g] (70–8610)	12.2[g] (1.0–35.9) 4.9[g] (0.4–13.1)	1.59[gc] (0.06–140.60) 0.15[gc] (0.01–5.04)	yes	yes	yes[##]	yes[##]	Δ Mn-U greater than Δ Mn-B	Roels et al. 1987
67 ore mill workers and battery installers 58 controls	4180	22.6 (4.0–49.0) 13.0 (1.0–46.0)	5.97 (1.00–110.50) 1.73 (1.00–15.00)	yes	yes	yes	yes		Gan et al. 1988
92 battery installers 101 controls	948[g] (46–10480)	8.1[g] (2.1–21.0) 6.8[g] (2.5–13.1)	0.84[g] (0.15–7.33) 0.09[g] (0.01–0.49)	yes	yes	no[##]	no[##]		Roels et al. 1992
10 welders 65/154 controls	364 (80–690)	16.5[g] (11.5–21.4) 10.4[g] (6.0–14.3)	1.07[gc] (0.61–1.86) 0.31[gc] (0.10–0.93)	yes	yes	yes[##]	yes[##]	65 controls Mn-B 154 controls Mn-U	Järvisalo et al. 1992
39 battery installers 29 smelters 35 controls	950[g] (50–10840) 1370[g] (60–17010)		0.51[gc] (0.1–10.0) 31.6[gc] (3.1–301.6) 0.17[gc] (0.04–1.30)	–	yes	–	–	Mn bioavailability from salts better than from MnO_2	Buchet et al. 1993
115 foundry workers 145 controls	225[g] (14–11480)[#]	10.3[g] (7.9–13.1)[#] 6.8[g] (5.3–8.6)[#]	0.73[gc] (0.47–0.96)[#] 0.62[gc] (0.39–0.99)[#]	yes	no	–	–	Mn-B: 69 matched pairs Mn-U: 56 matched pairs	Mergler et al. 1994
19 office workers 19 foundry workers 20 foundry workers	27[g] (9–40) 124[g] (72–620) 270[g] (120–650)	6.0[g] (4.0–7.4) 8.6[g] (7.9–9.5) 11.9[g] (9.6–18.0)	1.7[g] (0.7–7.0) 2.3[g] (0.7–6) 2.8[g] (0.9–5)	yes	yes	yes[##]	yes[##]	mean exposure-free interval 13 days (1–42)	Lucchini et al. 1995
123 foundry workers 27 controls	167 (6–1628)	10 (1–74) 7 (3–16)	2.1 (0.2–23.1) 0.2 (0.1–4.1)	yes	yes	weak[##]	no[##]		Maroni 1997
35 foundry workers 37 controls	193[g] (26–750)	9.8[g] (4.6–23.4) 6.8[g] (4.8–10.9)	3.0[g] (0.5–23.0) 0.4[g] (0.1–2.0)	yes	yes	yes[##]	no	Mn-A: unclear wether 93 or 193 $\mu g/m^3$	Lucchini et al. 1997

Table 9. continued

Groups	Mn in air (µg/m³)	Mn in blood (µg/l)	Mn in urine (µg/l)c	Δ Mn-B (exposed/ controls)	Δ Mn-U (exposed/ controls)	Correlation Mn-A to Mn-B	Correlation Mn-A to Mn-U	Remarks	Authors
39 battery installers	4 (1–12)	10.7 (3.9–25.8)	0.26 (0.10–1.80)	yes	no	yes	no	mean exposure-free interval 2 days	Bader et al. 1999
22 battery installers	37 (12–64)	11.7 (3.2–23.0)	0.33 (0.10–1.30)						
39 battery installers	387 (137–794)	13.8 (6.1–23.3)	0.49 (0.10–2.20)						
17 controls		7.5 (2.6–15.1)	0.39 (0.10–1.20)						
57 foundry workers	54 (5–1490)	9.7 (4.0–19.0)	1.81 (0.30–5.00)	yes	yes	no[##]	no[##]		Lucchini et al. 1999
87 controls		6.0 (2.0–9.5)	0.67 (0.06–5.00)						
13 foundry workers	39 (7–64)	13.1 (8.1–20.4)	–	yes	–	–	–	Mn-B: significant reduction after 3–4 weeks exposure-free interval	Lander et al. 1999
7 foundry workers	5 (2–8)	9.2 (7.5–11.2)							
4 foundry workers	–	16.8 (13.9–25.1)							
22 recycling workers	6 (4–8)	8.8 (5.0–19.5)							
90 controls	–	8.7 (4.1–19.0)							

Mn-A: manganese in the air (inhalable dust fraction); Mn-B: manganese in blood; Mn-U: manganese in urine; g: geometric mean; m: median; Δ: significant difference between the exposed persons and controls ($p<0{,}05$); c: related to creatinine; #: 25th/75th percentile; ##: correlation with long-term exposure index

104   *Manganese and its inorganic compounds*

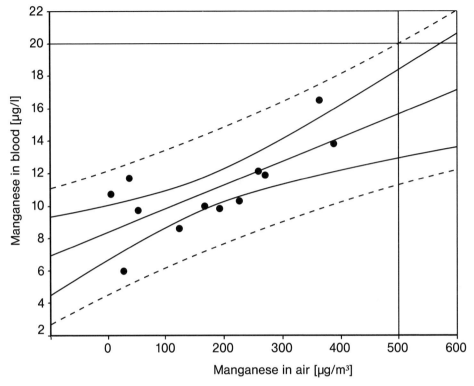

Figure 1. Mean values of the manganese concentrations in air and blood (Bader *et al*. 1999, Järvisalo *et al*. 1992, Lucchini *et al*. 1995, 1999, Maroni 1997, Mergler *et al*. 1994, Zschiesche *et al*. 1986). Shown are the compensating curves (middle line) and the 90 % confidence interval of the compensating curves (dotted line) and of the individual values (dashed line) (linear regression: $y = 1.46\ \mu g/l \times (100\ \mu g/m^3)^{-1} + 8.36\ \mu g/l$).

On the basis of linear regression analysis, a mean level of manganese in blood of 16 µg/l was determined with a manganese concentration of 0.5 mg/m³ (MAK value) in air. The 95 % confidence interval for the individual values with exposure at the level of the MAK value yielded a concentration of about 20 µg/l.

Figure 2 shows the results of the determination of manganese in urine. Creatinine-related mean values were multiplied by a factor of 1.5 to allow better comparability. In addition, the studies of Mergler *et al.* (1994) and Bader *et al.* (1999) were not included in the evaluation, as no differences between the exposed groups and controls were determined.

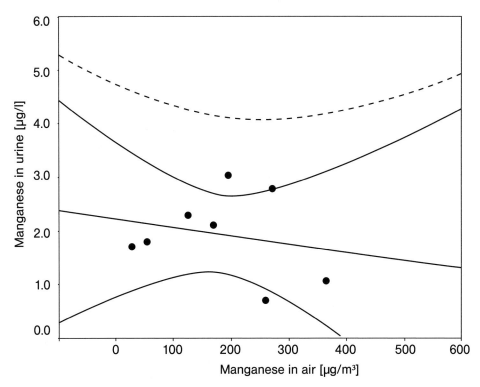

Figure 2. Mean values of the manganese concentrations in air and urine (Järvisalo *et al.* 1992, Lucchini *et al.* 1995, 1999, Maroni 1997, Zschiesche *et al.* 1986). Shown are the compensating curves (middle line) and the 90 % confidence interval of the compensating curves (dotted line) and of the individual values (dashed line) (linear regression: y = 0.015 µg/l × (100 µg/m³)$^{-1}$ + 2.2 µg/l).

The relationship between the manganese concentrations in air and urine is not statistically significant (p = 0.617).

The correlation coefficient for manganese in blood and manganese in urine is r = –0.368 (p = 0.370; product–moment correlation according to Pearson). This non-significant result confirms the observation reported in all the studies evaluated that there is no relevant relationship between the manganese concentrations in blood and urine. A BAT value for manganese in urine cannot therefore be evaluated on the basis of the relationship between external and internal exposure.

5 Background Exposure

5.1 Manganese in blood

The manganese was usually determined in blood samples taken at the end of a workshift. Exceptions to this were the studies of Lucchini et al. 1995 (on average after 13 exposure-free days) and of Bader et al. 1999 (on average after 2 exposure-free days).

The determination was carried out in all cases by means of atomic absorption spectrometry. Possible contamination of the biological material as a result of dust containing manganese (e.g. from working clothes) or from manganese steel needles, and measures for avoiding this are explicitly discussed by several authors (Bader et al. 1999, Lucchini et al. 1995, Mergler et al. 1994, Roels et al. 1987).

The results for the manganese concentrations in the blood of the control groups not occupationally exposed agree with data for the background exposure of the population (see Table 10).

The highest manganese concentrations in the blood of control persons were reported to be 13.1 µg/l by Roels et al. (1987), 16 µg/l by Maroni (1997) and 15.1 µg/l by Bader et al. (1999). An exception was the study of Gan et al. (1988), in which manganese levels of up to 49 µg/l blood were found. The mean manganese concentrations in the blood of the control persons varied between 4.9 µg/l (Roels et al. 1987) and 10.4 µg/l (Järvisalo et al. 1992). In most studies an average value of about 7 µg/l was given.

Table 10. Background exposure to manganese in whole blood and urine

Region[1]	Number of persons	Manganese in blood (µg/l)	Manganese in urine (µg/l)	Authors
EU	20	12.2	0.65	Buchet et al. 1976
EU	81	5.7 (0.4–13.1)	0.27 (0.1–2.0)	Lauwerys et al. 1985
EU	60	10	0.6	Kessel et al. 1987
EU	341/79	–	0.9/1.1	Alessio et al. 1989
EU	62	15.8	–	Kondakis et al. 1989
EU	88/777	8.8 (5.0–12.4)	1.02 (0.12–1.9)	Minoia et al. 1990
EU	36	6.8 (5.0–11.0)	0.43 (0.1–2.0)	Lucchini et al. 1994
EU	15	6.2 (4.8–8.4)	0.43 (0.1–2.0)	Smargiassi et al. 1995
EU	27	–	0.26	Jiménez-Jiménez et al. 1995
NA	60	9.1 (3.9–15.1)	–	Pleban and Pearson 1979
NA	10	–	0.4 c	Greger et al. 1990
NA	30	6.7 (2.0–13.1)	–	Loranger and Zayed 1995
SA	25	–	4.0 (0.4–8.1)	Siqueira and Moraes 1989
AS	232	34.7	–	Horiuchi et al. 1967
AS	16	–	4.8	Foo et al. 1993
AS	17	23.3 (17.3–30.1)	3.9 (0.7–9.6)	Chia et al. 1993
AS	98/44	17.8	1.16	Kaji et al. 1993
AU	15	11.8	–	Hams and Fabri 1988

[1] EU = Europe, NA = North America, SA = South America, AS = Asia, AU = Australia
c = related to creatinine

Two conclusions can be drawn about the manganese concentrations in the blood of persons not occupationally exposed:
1. The background exposure in the European and North American regions is lower than in Asia.
2. In Europe and North America the values vary for persons not occupationally exposed to manganese from between 6 µg/l (e.g. Roels *et al.* 1987) and about 16 µg/l (Kondakis *et al.* 1989). In four studies from the East Asian region mean concentrations of between 13 µg/l (Gan *et al.* 1988) and 35 µg/l (Horiuchi *et al.* 1967) were reported. This difference in the background exposure is possibly the result of nutritional and geological factors, i.e. a higher level of manganese in the soil and thus in food and drinking water (Alessio and Lucchini 1997).

In more recent studies lower manganese concentrations in biological materials were reported than before 1991.

For 618 persons from the 10 European studies since 1990 (see Tables 9 and 10) a weighted mean value of about 7.4 µg/l was calculated. For the period from 1971 to 1989 (142 test persons from 3 studies) a mean concentration of 12.8 µg/l can be determined. An exception are the results of Roels *et al.* (1987) and of Lauwerys *et al.* (1985). Mean manganese concentrations in blood of below 6 µg/l were reported by these authors as early as 1985. According to Valentin and Schiele (1980) this could be the result of an improvement in the analytical specificity of the atomic absorption spectrometric determination of manganese.

From the data for background exposure of all the more recent studies named in Table 9 and Table 10 a mean manganese level in blood of 7–8 µg/l can therefore be calculated for the European region. In most cases a concentration of about 10–12 µg/l is given as the upper reference value (95th percentile).

5.2 Manganese in urine

In the urine of the reference groups investigated, mean manganese concentrations in the range between 0.1 µg/g creatinine (Roels *et al.* 1992) and 0.7 µg/l urine (Lucchini *et al.* 1999) were found. In most cases renal excretion was about 0.2 to 0.3 µg/l.

As regards the background exposure of the general population, the levels of manganese in urine, unlike manganese in blood, do not indicate geographical or temporal trends. The results of studies of environmental exposure to manganese vary in the range between 0.2 µg/l and 2 µg/l (Table 10). Mean values of about 0.5 µg/l were given by several authors (e.g. Hochberg *et al.* 1996, Lucchini *et al.* 1995, Smargiassi *et al.* 1995), while in other studies mean concentrations of about 1 µg/l were reported (e.g. Kaji *et al.* 1993, Mergler *et al.* 1994, Minoia *et al.* 1990). The upper reference value in all studies was about 2 µg/l. This value was frequently exceeded, with values up to 5 µg/l, above all in Asian and South American regions (Chia *et al.* 1993, Foo *et al.* 1993, Siqueira and Moraes 1989).

6 Selection of the Indicators

The parameters suitable for the biological monitoring of exposed workers are primarily the levels of manganese in blood and in urine (with the restrictions already mentioned).

6.1 Manganese in blood

As manganese is found bound to erythrocytes and plasma proteins, the investigation medium necessary is whole blood.

The level of manganese in blood reflects the total exposure of the organism (Alessio and Lucchini 1997). After pulmonary or gastrointestinal absorption, the metal is rapidly distributed in the body tissues. Manganese is eliminated with a half-time of about 36 to 41 days. The manganese concentration in blood is therefore proportional to the total amount absorbed over a period of several weeks. Although special requirements regarding the sampling time are therefore not necessary, for the sake of uniformity sampling should be carried out at the end of the shift.

During sampling care must be taken that the materials used are free from contamination.

In ten of twelve studies the manganese concentrations in the blood of the exposed workers on a group basis differ significantly from those in the blood of the controls. In one of the other two studies no control group was investigated (Zschiesche et al. 1986), and in the other study biological monitoring was not carried out in blood (Buchet et al. 1993).

A possible relationship between external and internal exposure was investigated in ten studies (Bader et al. 1999, Gan et al. 1988, Järvisalo et al. 1992, Lucchini et al. 1995, 1997, 1999, Maroni 1997, Roels et al. 1987, 1992, Zschiesche et al. 1986). All authors concluded that there is no relationship between the current manganese concentration in the air and that in blood. This applied both for studies in which only personal air sampling was carried out and for the results of static sampling.

Bader et al. (1999) found in three subgroups with different levels of exposure a statistically significant increase in the manganese concentrations in blood after moderate external exposure ($p < 0.1$; Spearman's rank correlation).

The authors of the studies named above interpreted this result as evidence that manganese in blood primarily reflects the total exposure of the body. As a result of the rapid distribution of manganese out of the blood and into body tissue, short-term changes in exposure do not immediately lead to corresponding changes in the level of manganese in blood.

To take into account previous exposure, in seven studies so-called chronic exposure indices (CEIs) were formed. On the basis of the current and previous exposure to manganese in the air, the duration of work of the employee under these conditions and the personal estimations of the employee or the investigator, the CEI allows the differentiated consideration of the total exposure.

In three studies a statistically significant correlation between the CEI and the manganese concentration in blood was determined ($p < 0.05$) (Lucchini et al. 1995,

1997, Roels *et al.* 1987). Maroni (1997) also found a relationship between the two parameters, which, however, was not statistically significant (p = 0.16).

6.2 Manganese in urine

The manganese concentrations in urine samples were determined, like those in the blood samples, by means of atomic absorption spectrometry. Sampling was carried out as a rule at the end of the workshift.

The mean manganese concentrations in the urine of the exposed persons varied in the range between 0.3 µg/l (Bader *et al.* 1999) and 5.8 µg/l (Gan *et al.* 1988). The highest exposure levels of 140.6 µg/g creatinine and 301.6 µg/g creatinine were reported by Roels *et al.* (1987) and Buchet *et al.* (1993). In most of the other studies the manganese concentrations in urine reached levels of about 5 to 10 µg/l or per gram creatinine. In several studies the lowest given manganese concentration in urine corresponded with the detection limit. This observation suggests that manganese was not found in all samples. The number of urine samples in which no manganese was detected was given, however, only by Bader *et al.* (1999). These authors found manganese in the urine of employees with low-level exposure only in about 30 % of cases, with moderate exposure in about 50 % and with higher levels of exposure in 75 % of cases.

The manganese eliminated with the urine primarily reflects current exposure (Alessio and Lucchini 1997, Andersen *et al.* 1999). To avoid contamination of the urine samples, care must be taken that the samples are not given in working clothes and only after washing the hands.

7 Methods

For the analysis of manganese in urine there are two methods available that have been tested and recommended by the working group "Analyses of Hazardous Substances in Biological Materials" of the DFG Commission for the Investigation of Health Hazards of Chemical Compounds in the Work Area (Angerer and Schaller 1982, 2001). For the determination of the metal in blood a procedure is given which determines the manganese by means of flameless atomic absorption spectrometry (AAS) (Bader *et al.* 1999). The method has been tested and approved by the DFG working group "Analyses of Hazardous Substances in Biological Materials".

The detection limit of the atomic absorption spectrometric method is 0.5 µg/l in urine (Angerer and Schaller 1982). The in-series precision is between 2 % for a concentration of about 10 µg/l and 6 % for a concentration of about 5 µg/l. This urinary analysis based on high-resolution ICP-MS has a detection limit of 10 ng/l urine (Angerer and Schaller 2001). In blood Bader *et al.* (1999) determined a detection limit of 0.5 µg/l. The in-series precision in this case was 5.3 % (manganese concentration: 2.5 µg/l), the day-to-day precision 10.4 %. In external quality control programmes the variance of the method in the range of 20 µg/l was found to be about 7 %. Exposure at the level of the suggested BAT value can therefore be reliably distinguished from the background exposure (10–12 µg/l).

8 Evaluation of the BAT Value

To evaluate a BAT value, both the relationship between internal exposure and adverse effects on the health and also the correlation between external exposure (air) and internal exposure (blood) can be used.

Both approaches lead to more or less the same result, although the results for the relationship between internal exposure to manganese and adverse effects on health are of greater importance for evaluating a threshold value.

On a statistical basis differences in various target variables (tremor, short-term memory, reaction time and tapping) are already detected in the concentration range of 10 to 15 µg/l blood. Meta-analysis of the available data, however, does not reveal any dose-response relationship. In addition, the findings are inconsistent in some cases and their clinical relevance unclear. As a result of these limitations, a BAT value cannot be deduced on the basis of the studies included in the meta-analysis. An adequate basis for the evaluation of a BAT value is the relationship discussed in Section 4.3 between the external exposure in the air and the manganese concentration in blood (Figure 1). Linear regression analysis yielded a mean level of manganese in blood of 16 µg/l for a manganese concentration of 0.5 mg/m^3 in the air. As the BAT value is conceived as the ceiling concentration for an individual, the 95 % confidence interval should be used with exposure at the level of the MAK value for setting the BAT value. This is about 20 µg/l blood.

This value differs by a factor of about 2 from the background exposures usually found in Europe (Table 10).

On the basis of the statistical evaluation of the available data for external and internal exposure to manganese of exposed employees a BAT value has therefore been set of

20 µg manganese per l blood

As a result of the toxicokinetic properties of the substance, sampling must be carried out after the end of exposure or the end of the shift, and with long-term exposure after several previous shifts.

9 References

Alessio L, Apostoli P, Ferioli A, Lombardi S (1989) Interference of manganese on neuroendocrinal system in exposed workers. Biol Trace Elem Res 21: 249–253

Alessio L, Lucchini R (1997) Manganese and manganese compounds. in: Argentesi F, Roi R, Marcos JMS (Eds) Data profiles for selected chemicals and pharmaceuticals. European Commission, Joint Research Centre, Ispra, Italy, 45–147

Amr M, Allam M, Osmaan AL, el-Batanouni M, el Samra G, Halim Z (1993) Neurobehavioral changes among workers in some chemical industries in Egypt. Environ Res 63: 295–300

Andersen ME, Gearhart JM, Clewell HJ (1999) Pharmacokinetic data needs to support risk assessments for inhaled and ingested manganese. Neurotoxicology 20: 161–172

Angerer J, Schaller KH (1982) Mangan. Bestimmung im Harn. in: Senatskommission zur Prüfung gesundheitsschädlicher Arbeitsstoffe (Ed.) Analytische Methoden zur Prüfung gesundheitsschädlicher Arbeitsstoffe. Band 2: Analysen in biologischem Material. 6. Lieferung, Wiley-VCH, Weinheim

Angerer J, Schaller KH (2001) Aluminium, chromium, cobalt, copper, manganese, molybdenum, nickel, vanadium. Determination in urine. in: Greim H (Ed.) Analyses of hazardous substances in biological materials, Vol. 7, Wiley-VCH, Weinheim

Araki S, Aono H, Yokoyama K, Murata K (1986) Filterable plasma concentration, glomerular filtration, tubular balance, and renal clearance of heavy metals and organic substances in metal workers. Arch Environ Health 41: 216–221

Aschner M, Vrana KE, Zheng W (1999) Manganese uptake and distribution in the central nervous system (CNS). Neurotoxicology 20: 173–180

Baader E (1954) Gewerbekrankheiten. Klinische Grundlagen der 40 meldepflichtigen Berufskrankheiten. Urban & Schwarzenberg Verlag, München–Berlin, 86–94

Bader M, Dietz MC, Ihrig A, Triebig G (1999) Biomonitoring of manganese in blood, urine and axillary hair following low-dose exposure during the manufacture of dry cell batteries. Int Arch Occup Environ Health 72: 521–527

Barceloux DG (1999) Manganese. Clin Toxicol 37: 293–307

Beuter A, Mergler D, de Geoffroy A, Carrière L, Bélanger S, Varghese I, Sreekumar J, Gauthier S (1994) Diadochokinesimetry: a study of patients with Parkinson's disease and manganese exposed workers. Neurotoxicology 15: 655–664

Buchet JP, Lauwerys R, Roels H, De Vos C (1976) Determination of manganese in blood and in urine by flameless atomic absorption spectrophotometry. Clin Chim Acta 73: 481–486

Buchet JP, Magos C, Roels H, Ceulemans E, Lauwerys R (1993) Urinary excretion of homovanillic acid in workers exposed to manganese. Int Arch Occup Environ Health 65: 131–133

Chandra SV, Skukla GS, Srivastava RS, Singh H, Gupta VP (1981) An exploratory study of manganese exposure to welders. Clin Toxicol 18: 407–416

Chia SE, Foo SC, Gan SL, Jeyaratnam J, Tian CS (1993) Neurobehavioral functions among workers exposed to manganese ore. Scand J Work Environ Health 19: 264–270

Chu NS, Hochberg FH, Calne DB, Olanow CW (1995) Neurotoxicology of manganese. in: Chang, LW, Dyer RS (Eds) Handbook of neurotoxicology. Marcel Dekker, New York–Basel–Hongkong, 91–103

Cook DG, Fahn S, Brait KA (1974) Chronic manganese intoxication. Arch Neurol 30: 59–64

Cotzias GC, Horiuchi K, Fuenzalida S, Mena I (1968) Chronic manganese poisoning: clearance of tissue manganese concentrations with persistence of the neurological pictures. Neurology18: 376–382

Couper J (1837) On the effect of black oxide of manganese when inhaled into the lungs. Br Ann Med Pharm Vital Stat Gen Sci 1: 41–42

Crump KS, Rousseau P (1999) Results from eleven years of neurological health surveillance at a manganese oxide and salt producing plant. Neurotoxicology 20: 273–286

Dietz MC, Schunk W (1998) Mangan. in: Triebig G, Lehnert G (Eds): Neurotoxikologie in der Arbeitsmedizin und Umweltmedizin. Gentner Verlag, Stuttgart, 333–343

Dietz MC, Wrazidlo W, Ihrig A, Bader M, Triebig G (2000) Magnetresonanztomographie des Gehirns bei Beschäftigten mit chronischer beruflicher Mangandioxid-Exposition. Rofo Fortschr Geb Rontgenstr Neuen Bildgeb Verfahr 172: 514–520

Dietz MC, Ihrig A, Bader M, Triebig G (2003) Arbeitsmedizinische Feldstudie zur chronischen Neurotoxizität von Mangandioxid. Arbeitsmed Sozialmed Umweltmed 38: 57–66

Emara AM, el-Ghawabi SH, Madkour OI, el-Samra GH (1971) Chronic manganese poisoning in the dry battery industry. Br J Ind Med 28: 78–82

Ferraz HB, Bertolucci PH, Pereira JS, Lima JG, Andrade LA (1988) Chronic exposure to the fungicide maneb may produce symptoms and signs of CNS manganese intoxication. Neurology 38: 550–553

Foo SC, Khoo NY, Heng A, Chua LH, Chia SE, Ong CN, Ngim CH, Jeyaratnam J (1993) Metals in hair as biological indices for exposure. Int Arch Occup Environ Health 65 (1 Suppl): S83–86

Fricke R, Treinies G (1985) Einführung in die Metaanalyse. Hans Huber, Bern

Gan SI, Tan KT, Kwok SF (1988) Biological threshold limit values for manganese dust exposure. Singapore Med J 29: 105–109

Geigy (1985) Wissenschaftliche Tabellen. Teilband Körperflüssigkeiten. 8. Auflage, 4. Nachdruck, Basel, 62

Gibbs JP, Crump KS, Houck DP, Warren PA, Mosley WS (1999) Focused medical surveillance: A search for subclinical movement disorders in a cohort of U.S. workers exposed to low levels of manganese dust. Neurotoxicology 20: 299–314

Greger JL, Davis CD, Suttie JW, Lyle BJ (1990) Intake, serum concentrations, and urinary excretion of manganese by adult males. Am J Clin Nutr 51: 457–461

Greim H (Ed.) (1999) Manganese and its inorganic compounds. in: Occupational toxicants, Vol.12, Wiley-VCH, Weinheim

Hams GA, Fabri JK (1988) Analysis for blood manganese used to assess environmental exposure. Clin Chem 34: 1121–1123

Hochberg F, Miller G, Valenzuela R, McNelis S, Crump KS, Covington T, Valdivia G, Hochberg B, Trustman JW (1996) Late motor deficits of Chilean manganese miners: a blinded control study. Neurology 47: 788–795

Horiuchi K, Horiguchi S, Tanaka N, Shinagawa K, Hamaguchi T (1967) Manganese contents in the whole blood, urine and feces of a healthy Japanese population. Osaka City Med J 13: 151–163

Hua MS, Huang CC (1991) Chronic occupational exposure to manganese and neurobehavioral function. J Clin Exp Neuropsychol 13: 495–507

Iregren A (1990) Psychological test performance in foundry workers exposed to low levels of manganese. Neurotoxicol Teratol 12: 673–675

Iregren A (1994) Using psychological tests for the early detection of neurotoxic effects of low level manganese exposure. Neurotoxicology 15: 671–677

Iregren A (1999) Manganese neurotoxicity in industrial exposures: proof of effects, critical exposure level, and sensitive tests. Neurotoxicology 20: 315–323

Järvisalo J, Olkinuora M, Kiilunen M, Kivistö H, Ristola P, Tossavainen A, Aitio A (1992) Urinary and blood manganese in occupationally nonexposed populations and in manual metal arc welders of mild steel. Int Arch Occup Environ Health 63: 495–501

Jiménez-Jiménez F, Molina J, Aguilar M, Arrieta F, Jorge-Santamaria A, Cabrera-Valdivia F, Ayuso-Peralta L, Rabasa M, Vazquez A, Garcia-Albea E, Martinez-Para M (1995) Serum and urinary manganese levels in patients with Parkinson's disease. Acta Neuro Scand 91: 317–320

Kaji H, Ohsaki Y, Rokujo C, Higashi T, Fujino A, Kamada T (1993) Determination of blood and urine manganese (Mn) concentrations and the application of static sensography as the indices of Mn-exposure among Mn-refinery workers. JUOEH 15: 287–296

Kessel R, Benzce K, Tewordt M, Mauermayer R, Friesen A (1987) Untersuchungen zum Mangangehalt in menschlichen Geweben. in: Norpoth K (Ed.) Dokumentationsband über die 27. Jahrestagung der Deutschen Gesellschaft für Arbeitsmedizin und Umweltmedizin e.V., Gentner-Verlag Stuttgart, 489–491

Koelsch F (Ed.) (1963) Lehrbuch der Arbeitsmedizin, Band I. Enke Verlag, Stuttgart

Kondakis XG, Makris N, Leotsinidis M, Prinou M, Papapetropoulos T (1989) Possible health effects of high manganese concentration in drinking water. Arch Environ Health 44: 175–178

Lander F, Kristiansen J, Lauritsen JM (1999) Manganese exposure in foundry furnacemen and scrap recycling workers. Int Arch Occup Environ Health 72: 546–550

Larsen NA, Pakkenberg H, Damsgaard E, Heydorn K (1979) Topographical distribution of arsenic, manganese, and selenium in the normal human brain. J Neurol Sci 42: 407–416

Lauwerys R, Roels H, Genet P, Toussaint G, Bouckaert A, De Cooman S (1985) Fertility of male workers exposed to mercury vapor or to manganese dust: a questionnaire study. Am J Ind Med 7: 171–176

Lehrl S, Gallwitz A, Blaha L, Fischer B (1992) Geistige Leistungsfähigkeit–Theorie und Messung der biologischen Intelligenz mit dem Kurztest KAI. 3. Auflage, Vless Verlag, Ebersberg

Loranger S, Zayed J (1995) Environmental and occupational exposure to manganese: a multimedia assessment. Int Arch Occup Environ Health 67: 101–110

Lucchini R, Smargiassi A, Bergamaschi E, Apostoli P (1994) Olfactory function and hand steadiness in manganese exposed workers. in: Fifth International Symposium on Neurobehavioral Methods and Effects in Occupational and Environmental Health (cited from Alessio and Lucchini 1997)

Lucchini R, Selis L, Folli D, Apostoli P, Mutti A, Vanoni O, Iregren A, Alessio L (1995) Neurobehavioral effects of manganese in workers from a ferroalloy plant after temporary cessation of exposure. Scand J Work Environ Health 21: 143–149

Lucchini R, Bergamaschi E, Smargiassi A, Festa D, Apostoli P (1997) Motor function, olfactory threshold, and hematological indices in manganese-exposed ferroalloy workers. Environ Res 73: 175–180

Lucchini R, Apostoli P, Perrone C, Placidi D, Albini E, Migliorati P, Mergler D, Sassine MP, Palmi S, Alessio L (1999) Long term exposure to "low levels" of manganese oxides and neurofunctional changes in ferroalloy workers. Neurotoxicology 20: 287–297

Maroni M (Ed.) (1997) Study of toxic effects on the central and peripheral nervous system of workers in the ferro-alloy industry. 5th ECSC medical research programme, Cat. number CE-V/4-91-006-EN-C, European Community, Brussels

McMillan DE (1999) A brief history of the neurobehavioral toxicity of manganese: some unanswered questions. Neurotoxicology 20: 499–508

Mena I, Marin O, Fuenzalida S, Cotzias GC (1967) Chronic manganese poisoning: clinical picture and manganese turnover. Neurology 17: 128–136

Mena I, Horiuchi K, Burke K, Cotzias GC (1969) Chronic manganese poisoning. Individual susceptibility and absorption of iron. Neurology 17: 1000–1006

Mergler D, Huel G, Bowler R, Iregren A, Bélanger S, Baldwin M, Tardif R, Smargiassi A, Martin L (1994) Nervous system dysfunction among workers with long-term exposure to manganese. Environ Res 64: 151–180

Mergler D, Baldwin M (1997) Early manifestations of manganese neurotoxicity in humans: An update. Environ Res 73: 92–100

Mergler D, Bélanger S, Larribe F, Panisset M, Bowler R, Baldwin M, Lebel J, Hudnell K (1998) Preliminary evidence of neurotoxicity associated with eating fish from the upper St. Lawrence River Lakes. Neurotoxicology 19: 691–702

Minoia C, Sabbioni E, Apostoli P, Pietra R, Pozzoli L, Gallorini M, Nicolaou G, Alessio L, Capodaglio E (1990) Trace element reference values in tissues from inhabitants of the European Community. I. A study of 46 elements in urine, blood and serum of Italian subjects. Sci Tot Environ 95: 89–105

Newland MC, Cox C, Hamada R, Oberdoerster G, Weiss B (1987) The clearance of manganese chloride in the primate. Fundam Appl Toxicol (9): 314–328

Oberdoerster G, Cherian G (1988) Manganese. in: Clarkson TW, Friebed L, Nordberg GF, Sager PR (Eds) Biological monitoring of toxic metals. Plenum Press, New York, 283–301

Pal PK, Samii A, Calne DB (1999) Manganese neurotoxicity: a review of clinical features, imaging and pathology. Neurotoxicology 20: 227–238

Pflaumbaum W, Blome H, Heidermanns G (1990) Vorkommen und Messung von Mangan und Manganverbindungen am Arbeitsplatz. Staub–Reinhaltung der Luft 50: 307–310

Pleban PA, Pearson KH (1979) Determination of manganese in whole blood and serum. Clin Chem 25: 1915–1918

Rodier J (1955) Manganese poisoning in Moroccan miners. Brit J Ind Med 12: 21–35

Roels H, Lauwerys R, Genet P, Sarhan MJ, de Fays M, Hanotiau I, Buchet JP (1987) Relationship between external and internal parameters of exposure to manganese in workers from a manganese oxide and salt producing plant. Am J Ind. Med 11: 297–305

Roels HA, Ghyselen P, Buchet JP, Ceulemans E, Lauwerys RR (1992) Assessment of the permissible exposure level to manganese in workers exposed to manganese dioxide dust. Br J Ind Med 49: 25–34

Roels HA, Ortega Eslava MI, Ceulemans E, Robert A, Lison D (1999) Prospective study on the reversibility of neurobehavioral effects in workers exposed to manganese dioxide. Neurotoxicology 20: 255–271

Seeber A, Dotzauer H, Schneider H (1984) Psychodiagnostik bei Blei-, Quecksilber- und Manganexposition – ein Erfahrungsbericht. Z Gesamte Hyg 30: 702–706

Siegl P, Bergert KD (1982) Eine frühdiagnostische Überwachungsmethode bei Manganexponierten. Z Gesamte Hyg 28: 524–526

Siqueira ME, Moraes EC (1989) Homovanillic acid (HVA) and manganese in urine of workers exposed in a ferromanganese alloy plant. Med Lav 80: 224–228

Sjögren B, Iregren A, Frech W, Hagman M, Johansson L, Tesarz M, Wennberg A (1996) Effects on the nervous system among welders exposed to aluminium and manganese. Occup Environ Med 53: 32–40

Smargiassi A, Mergler D, Bergamaschi E, Vettori MV, Lucchini R, Apostoli P (1995) Peripheral markers of catecholamine metabolism among workers occupationally exposed to manganese (Mn). Toxicol Lett 77: 329–333

Smeyers-Verbeke J, Bell P, Lowenthal A, Massart DL (1976) Distribution of Mn in human brain tissue. Clin Chim Acta 68: 343–347

Smyth LT, Ruhf RC, Whitman NE, Dugan T (1973) Clinical manganism and exposure to manganese in the production and processing of ferromanganese alloy. J Occup Med 15: 101–109

Tewes U (Ed.) (1991) Hamburg-Wechsler-Intelligenztest für Erwachsene, Revision 1991. Verlag Hans Huber, Bern

Triebig G (1998) Role of brain imaging techniques in occupational neurotoxicology. in: Costa LG, Manzo L (Eds): Occupational neurotoxicology. CRC Press, Boca Raton–Boston–London–New York–Washington DC, 199–211

Valentin H, Schiele R (1980) Human biological monitoring of industrial chemicals. Part 2: Manganese. Industrial Health and Safety 1980, Rapports EUR 6608, Commission of the European Community, Luxemburg

Wennberg A, Iregren A, Struwe G, Cizinsky G, Hagman M, Johansson L (1991) Manganese exposure in steel smelters a health hazard to the nervous system. Scand J Work Environ Health 17: 255–262

Wennberg A, Hagman M, Johansson L (1992) Preclinical neurophysiological signs of parkinsonism in occupational manganese exposure. Neurotoxicology 13: 271–274

WHO (World Health Organization) (1981) Manganese. Environmental Health Criteria 17, Geneva

WHO (World Health Organization) (1987) Diseases caused by manganese and its toxic compounds. in: WHO, Early detection of occupational diseases, Geneva, 69–73

Wolters EC, Huang CC, Clark C, Peppard RF, Okada J, Chu NS, Adam MJ, Ruth TJ, Li D, Calne DB (1989) Positron emmission tomography in manganese intoxication. Ann Neur 26: 647–651

Yiin SJ, Lin TH, Shih TS (1996) Lipid peroxidation in workers exposed to manganese. Scand J Work Environ Health 22: 381–386

Zidek W (1983) Mangan (Mn). in: Zumkley H (Ed.) Spurenelemente. Thieme Verlag, Stuttgart, 140–151

Zimmermann-Hölz HJ (1989) Mangan. in: Konietzko J, Dupuis H (Eds): Handbuch der Arbeitsmedizin. Ecomed Verlag, Landsberg–München–Zürich

Zschiesche W, Wilhelm E, Zober A, Schaller KH, Weltle D, Valentin H (1986) Manganese in stainless steel welding fumes – External exposure and biological monitoring. in: Stern RM, Berlin A, Fletcher AC, Järvisalo J (Eds). Health hazards and biological effects of welding fumes and gases. Excerpta Medica, Amsterdam–New York–Oxford, 193–196

Authors: G. Triebig, A. Ihrig and M. Bader
Approved by the Working Group: 20.02.2001

Tetrachloromethane, Addendum

BAT	3.5 µg tetrachloromethane/l whole blood Sampling time: end of exposure or end of shift, with long-term exposure after several consecutive shifts
Date of evaluation	2003

The toxicity of tetrachloromethane was originally reviewed in Volume 1 of the present series (Henschler and Lehnert 1994). In 2000 tetrachloromethane (carbon tetrachloride) was classified in Carcinogen category 4. The previously valid MAK value of 10 ml/m^3 was reduced to 0.5 ml/m^3 (Greim 2002). The BAT value has therefore been re-evaluated.

1 Evaluation of the BAT Value in Blood

Pharmacokinetic calculations showed that tetrachloromethane accumulates in adipose tissue in man, unlike in the rat (Paustenbach et al. 1988). They also showed that in man, after exposure to tetrachloromethane for 2 weeks (5 ml/m^3, 8 hours a day, 5 days a week, for 2 weeks), the maximum concentration of tetrachloromethane in venous blood is to be expected at the end of the second week at a level of about 35 µg/l. If one assumes that in this pharmacokinetic model a reduction in the level of exposure from 5 ml/m^3 to 0.5 ml/m^3 also leads to a reduction in the calculated concentration of tetrachloromethane in venous blood to one tenth of the above value, then for 8-hour exposure to tetrachloromethane concentrations of 0.5 ml/m^3, concentrations in whole blood of 3.5 µg/l are to be expected. This value corresponds with the findings of Brugnone et al. (1983).

The BAT value for tetrachloromethane in blood has therefore been set at

3.5 µg tetrachloromethane per litre whole blood

Sampling should be carried out at the end of the shift after several consecutive shifts.

In view of the very limited database, this BAT value should be regarded as provisional.

2 References

Brugnone F, Apostoli P, Perbellini L, Silvestri R, Cocheo V (1983) Monitoring of occupational exposure to low concentrations of carbon tetrachloride. in: Hayes AW, Schnell RC, Miya TS (Eds) Developments in the science and practice of toxicology: Proceedings of the Third International Congress of Toxicology held in San Diego, California, USA, August 28–September 3, 1983. Elsevier Science Publishers, Amsterdam–New York–Oxford, 575–578
Greim H (Ed.) (2002) Carbon tetrachloride. in: Occupational toxicants, Vol. 18, Wiley-VCH, Weinheim
Henschler D, Lehnert G (Eds) (1994) Tetrachloromethane. in: Biological exposure values for occupational toxicants and carcinogens – Critical data evaluation for BAT and EKA values, Vol. 1, VCH, Weinheim, pp. 153–161
Paustenbach DJ, Clewell HJ, Gargas ML, Andersen ME (1988) A physiologically based pharmacokinetic model for inhaled carbon tetrachloride. Toxicol Appl Pharmacol 96: 191–211

Author: H.M. Bolt
Approved by the Working Group: 05.05.2003

Tetrahydrofuran, Addendum

BAT	2.0 mg tetrahydrofuran/l urine
	Sampling time: end of exposure or end of shift
Date of evaluation	2001

The documentation for tetrahydrofuran was published in Volume 2 of this series (Greim and Lehnert 1995). Developments in the meantime have made re-evaluation of the BAT value necessary.

1 Evaluation of the BAT Value

The MAK value for tetrahydrofuran was lowered in 1996 from 200 ml/m^3 to 50 ml/m^3 as a result of morphological changes in the nasal mucosa of the rat (Greim 1996). The kidney and liver tumours observed in the NTP study (NTP 1998) led in 1999 to the classification of tetrahydrofuran in Carcinogen category 3. The MAK value was retained, as tetrahydrofuran is not genotoxic (Greim 1999). In 2002 the substance was designated with an "H" (for substances which can be absorbed in dangerous amounts through the skin) (Greim 2002) and in 2003 tetrahydrofuran was re-classified in Carcinogen category 4 (Greim 2003).

The BAT value for tetrahydrofuran of 8 mg/l urine set in 1993 therefore had to be re-evaluated in accordance with the lowered MAK value of 50 ml/m^3.

As a result of the limited database, the previous BAT value for tetrahydrofuran was evaluated exclusively from the relationship between external and internal exposure and was based on a Japanese field study of 58 workers (Ong *et al.* 1991). The correlation data were confirmed by a laboratory study with 8-hour exposure (see the BAT documentation for tetrahydrofuran in Volume 2 of this series; Greim and Lehnert 1995). On the basis of the data available and the resulting regression, for external exposure to a tetrahydrofuran concentration of 50 ml/m^3 the BAT value has been re-evaluated and set at

2.0 mg tetrahydrofuran per litre urine

Sampling should be carried out at the end of exposure or the end of the shift.
The given value is to be regarded as provisional.

2 References

Greim H, Lehnert G (Eds) (1995) Tetrahydrofuran. in: Biological exposure values for occupational toxicants and carcinogens – Critical data evaluation for BAT and EKA values, Vol. 2, VCH, Weinheim, pp. 99–105

Greim H (Ed.) (1996) Tetrahydrofuran. Gesundheitsschädliche Arbeitsstoffe, Toxikologisch-arbeitsmedizinische Begründungen von MAK-Werten, 23. Lieferung, VCH, Weinheim

Greim H (Ed.) (1999) Tetrahydrofuran. Gesundheitsschädliche Arbeitsstoffe, Toxikologisch-arbeitsmedizinische Begründungen von MAK-Werten, 28. Lieferung, Wiley-VCH, Weinheim

Greim H (Ed.) (2002) Tetrahydrofuran. Gesundheitsschädliche Arbeitsstoffe, Toxikologisch-arbeitsmedizinische Begründungen von MAK-Werten, 34. Lieferung, Wiley-VCH, Weinheim

Greim H (Ed.) (2003) Tetrahydrofuran. Gesundheitsschädliche Arbeitsstoffe, Toxikologisch-arbeitsmedizinische Begründungen von MAK-Werten, 35. Lieferung, Wiley-VCH, Weinheim

NTP (National Toxicology Program) (1998) Toxicology and carcinogenesis studies of tetrahydrofuran (CAS No. 109-99-9) in F344/N rats and B6C3F1 mice (inhalation studies). NTP Technical Report Series No. 475, NIH Publication No. 98-3965, US Department of Health and Human Services, National Institutes of Health, Bethesda, MD, USA

Ong CN, Chia SE, Phoon WH, Tan KT (1991) Biological monitoring of occupational exposure to tetrahydrofuran. Br J Ind Med 48: 616–621

Authors: J. Lewalter, G. Leng
Approved by the Working Group: 19.06.2000

Documentation for Carcinogenic Substances
With Biological Exposure Equivalents

Dichloromethane, Addendum

EKA

The correlations between external and internal exposure yield the following data:

Air Dichloromethane (ml/m³)	(mg/m³)	Whole blood Dichloromethane (mg/l)
10	35	0.1
20	70	0.2
50	175	0.5
100	350	1.0

Sampling time: during exposure, but at the earliest two hours after its commencement

Date of evaluation 2001

Since dichloromethane was reviewed in Volume 1 of this series (Henschler and Lehnert 1994), the substance has been classified on the basis of animal experiments as a carcinogen (Carcinogen category 3A) and the MAK value has been withdrawn (Greim 2000). The BAT values then valid were a carbon monoxide level in haemoglobin (COHb) of 5 % and a dichloromethane level in blood of 1 mg/l whole blood. As the reference value for the carbon monoxide load on the haemoglobin was removed with the withdrawal of the MAK value, re-evaluation of the BAT value became necessary.

1 Evaluation of the Biological Exposure Equivalents

1.1 Metabolism and metabolic activation

Dichloromethane is metabolized via a microsomal (oxidative) and a cytosolic (reductive) pathway. The oxidative pathway was found to be saturated in all the species investigated (mouse, rat, hamster and man) in the range of an exposure concentration of 500 ml/m³ (Greim 2000); this does not apply, however, for the reductive GSH-dependent pathway (Gargas *et al.* 1986).

It was confirmed that the metabolic pathway of dichloromethane mediated by glutathione *S*-transferase theta (GSTT1-1) is associated with the tumour-promoting effects of dichloromethane (summary: Green 1997); the chemical identity of the reactive metabolite formed via this pathway is very probably *S*-[chloromethyl]glutathione (Thier *et al.* 1993). The relationship between the reductive metabolic pathway of dichloromethane via GSTT1-1 and the formation of genotoxic and reactive metabolites was confirmed in other studies (Graves and Green 1996, Sherratt *et al.* 1998).

In man, *GSTT1* is polymorphically expressed. In about 20 % of the German population, neither the enzyme activity nor the gene for human GSTT1-1 (hGSTT1-1) can be detected (review: Thier *et al.* 2003). Because in these persons the conjugating activity of hGSTT1-1 is absent, they are called 'non-conjugators'; all other people are 'conjugators' (Hallier *et al.* 1990, Pemble *et al.* 1994, Peter *et al.* 1989). The conjugators are, however, not a homogeneous group but can be subdivided into two subgroups, namely low activity and high activity conjugators, on the basis of the enzyme activity determined in the erythrocytes and the number of *hGSTT1* alleles present (Hallier *et al.* 1990, Wiebel *et al.* 1999). In some other ethnic groups, the incidence of the non-conjugator genotype has been shown to be very different. For example, non-conjugators make up about 11 % of the Swedish population (Warholm *et al.* 1994), 60 % of the Chinese and Korean populations (Nelson *et al.* 1995, Shen *et al.* 1998), 38 % of Malays, 16 % in India (Lee *et al.* 1995) and in Americans of Mexican origin only 10 % (Nelson *et al.* 1995).

The effects of ethnic differences in the incidence of the homozygous *GSTT1* deletion on the estimates of risk from dichloromethane in man have been studied (El-Masri *et al.* 1999). For this purpose, Monte Carlo simulation methods were used in combination with an extended version of an existing toxicokinetic (PBPK, "physiologically based pharmacokinetics") model. Different calculations carried out for populations with different incidences of the homozygous deletion of the *hGSTT1* gene yielded marked differences in the risks estimated for these populations: the higher the proportion of individuals with homozygous gene deletions within a population, the lower was the risk caused by exposure to dichloromethane in this population.

1.2 Exposure equivalents for carcinogenic substances (EKA)

In the previous addendum to the BAT documentation for dichloromethane, studies were presented which describe the quantitative relationship between external exposure to dichloromethane and the level of dichloromethane in the blood in man; see Volume 1 of this series (Henschler and Lehnert 1994). With exposure to a constant dichloromethane concentration of 100 ml/m^3, a mean dichloromethane level in blood of 1 mg/l whole blood is established. These relationships led to the setting of the following correlations:

Air Dichloromethane (ml/m³)	(mg/m³)	Blood Dichloromethane (mg/l whole blood)
10	35	0.1
20	70	0.2
50	175	0.5
100	350	1.0

Sampling should be carried out during the exposure, but at the earliest two hours after its commencement.

2 References

El-Masri HA, Bell DA, Portier CJ (1999) Effects of glutathione transferase theta polymorphism on the risk estimates of dichloromethane to humans. Toxicol Appl Pharmacol 158: 221–230

Gargas ML, Clewell HJ, Andersen ME (1986) Metabolism of inhaled dihalomethanes *in vivo*: differentiation of kinetic constants for two independent pathways. Toxicol Appl Pharmacol 82: 211–223

Graves RJ, Green T (1996) Mouse liver glutathione *S*-transferase mediated metabolism of methylene chloride to a mutagen in the CHO/HPRT assay. Mutat Res 367: 143–150

Green T (1997) Methylene chloride induced mouse liver and lung tumours: an overview of the role of mechanistic studies in human safety assessment. Hum Exp Toxicol 16: 3–13

Greim H (Ed.) (2000) Dichloromethane. in: Occupational toxicants, Vol. 17, Wiley-VCH, Weinheim

Hallier E, Deutschmann S, Reichel C, Bolt HM, Peter H (1990) A comparative investigation of the metabolism of methyl bromide and methyl iodide in human erythrocytes. Int Arch Occup Environ Health 62: 221–225

Henschler D, Lehnert G (Eds) (1994) Dichloromethane. in: Biological exposure values for occupational toxicants and carcinogens – Critical data evaluation for BAT and EKA values, Vol. 1, VCH, Weinheim, pp. 45–55

Lee EJ, Wong JY, Yeoh PN, Gong NH (1995) Glutathione *S*-transferase-theta (*GSTT1*) genetic polymorphism among Chinese, Malays and Indians in Singapore. Pharmacogenetics 5: 332–334

Nelson HH, Wiencke JK, Christiani DC, Cheng TJ, Zuo ZF, Schwartz BS, Lee BK, Spitz MR, Wang M, Xu X (1995) Ethnic differences in the prevalence of the homozygous deleted genotype of glutathione *S*-transferase theta. Carcinogenesis 16: 1243–1245

Pemble SE, Schroeder KR, Spencer SR, Meyer DJ, Hallier E, Bolt HM, Ketterer B, Taylor JB (1994) Human glutathione *S*-transferase theta (*GSTT1*): cDNA cloning and the characterization of a genetic polymorphism. Biochem J 300: 271–276

Peter H, Deutschmann S, Reichel C, Hallier E (1989) Metabolism of methyl chloride by human erythrocytes. Arch Toxicol 63: 351–355

Shen JH, Lin GF, Yuan WX, Tan JW, Bolt HM, Thier R (1998) Glutathione transferase T1 and M1 genotype polymorphism in the normal population of Shanghai. Arch Toxicol 72: 456–458

Sherratt PJ, Manson MM, Thomson AM, Hissink EA, Neal GE, van Bladeren PJ, Green T, Hayes JD (1998) Increased bioactivation of dihaloalkanes in rat liver due to induction of class theta glutathione *S*-transferase T1-1. Biochem J 335: 619–630

Thier R, Taylor JB, Pemble SE, Humphreys WG, Persmark M, Ketterer B, Guengerich FP (1993) Expression of mammalian glutathione S-transferase 5-5 in *Salmonella typhimurium* TA1535 leads to base-pair mutations upon exposure to dihalomethanes. Proc Natl Acad Sci USA 90: 8576–8580

Thier R, Brüning T, Roos PH, Rihs HP, Golka K, Ko Y, Bolt HM (2003) Markers of genetic susceptibility in human environmental hygiene and toxicology: the role of selected CYP, NAT and GST genes. Int J Hyg Environ Health 206: 149–171

Warholm M, Alexandrie AK, Hogberg J, Sigvardsson K, Rannug A (1994) Polymorphic distribution of glutathione transferase activity with methyl chloride in human blood. Pharmacogenetics 4: 307–311

Wiebel FA, Dommermuth A, Thier R (1999) The hereditary transmission of the glutathione transferase hGSTT1-1 conjugator phenotype in a large family. Pharmacogenetics 9: 251–256

Author: H.M. Bolt
Approved by the Working Group: 20.02.2001

Tetrachloroethene, Addendum

EKA The correlations between external and internal exposure yield the following data:

Air Tetrachloroethene (ml/m^3)	(mg/m^3)	Whole blood Tetrachloroethene (mg/l)
10	69	0.2
20	138	0.4
30	206	0.6
50	344	1.0

Sampling time: about 16 hours after end of shift

Date of evaluation 2001

The toxicity of tetrachloroethene was reviewed in Volume 1 of the present series (Henschler and Lehnert 1994). After reclassification of tetrachloroethene in Carcinogen category 3B and withdrawal of the MAK value, the previous BAT value for tetrachloroethene of 1 mg/l blood (sampling 16 hours after the end of the shift) was reviewed. The substance has been reclassified as experiments yielded evidence that tetrachloroethene may cause cancer of the kidneys (Greim 1998).

1 Withdrawal of the BAT Value for Alveolar Air

The BAT values for tetrachloroethene evaluated in 1982 were given for both whole blood and alveolar air. In both cases sampling was recommended 16 hours after the end of exposure.

The quantitative determination of the tetrachloroethene concentration in whole blood sampled before exposure after several working days has proved reliable in practice, and the evaluated BAT value has been accepted both at a scientific level and by occupational physicians. Taking into consideration an external occupational exposure to tetrachloroethene of 50 ppm and the sampling conditions given above, the evaluated BAT value is in agreement with the American BEI.

A somewhat higher value of 1.6 mg/l blood was evaluated under the same conditions in a Korean study (Jang et al. 1993). The reason for this difference was thought to be e.g. ethnic differences (Jang et al. 1997).

The results of tetrachloroethene determinations in blood sampled before the shift for monitoring occupationally exposed persons, can therefore be reliably evaluated with the existing BAT value.

The determination of tetrachloroethene in alveolar air, however, has not become established in practical occupational medicine. This state of affairs is not substance-specific, but applies for determinations in alveolar air in general. This is mainly due to the fact that there are no practical and reliable sampling systems commercially available. Also the great influence on the values determined of the sampling time, the co-operation of the test person and analytical difficulties explain the rejection of this material for occupational-medical investigations. In view of this, the BAT value for the tetrachloroethene concentration in alveolar air has been withdrawn from the *List of MAK and BAT Values*.

2 Evaluation of the Biological Exposure Equivalents

As described in the chapter on tetrachloroethene in Volume 1 of this series (Henschler and Lehnert 1994), after absorption the substance is exhaled mainly unchanged in man. The metabolized amount is transformed mainly oxidatively into trichloroacetic acid, which is excreted in the urine. A quantitatively less important metabolic pathway, which, however, is associated with the occurrence of kidney tumours, takes place via the conjugation of tetrachloroethene with glutathione and transformation into the nephrotoxic metabolites via the "beta-lyase" of the kidneys (Völkel *et al.* 1998). Studies of the toxicokinetics and metabolism of tetrachloroethene led to the creation of new toxicokinetic models ("Physiologically Based Pharmacokinetics" (PBPK) and population models; Bois *et al.* 1996, Reitz *et al.* 1996).

The relationships between external exposure to tetrachloroethene and the levels of tetrachloroethene in blood are very well documented and supported by different independent studies (see the chapter on tetrachloroethene in Volume 1 of this series (Henschler and Lehnert 1994)). The preference of the tetrachloroethene level in blood for the biological monitoring of tetrachloroethene was particularly emphasised (Skender *et al.* 1991).

On the basis of the data for the previous BAT value, the following correlations between external exposure to tetrachloroethene and the levels of tetrachloroethene in blood can be established:

Air Tetrachloroethene		Blood Tetrachloroethene
(ml/m^3)	(mg/m^3)	(mg/l)
10	69	0.2
20	138	0.4
30	206	0.6
50	344	1.0

Sampling: about 16 hours after the end of the shift.
Otherwise, the information in the earlier chapter applies.

3 References

Bois FY, Gelman A, Jiang J, Maszle DR, Zeise L, Alexeef G (1996) Population toxicokinetics of tetrachloroethylene. Arch Toxicol 70: 347–355

Greim H (Ed.) (1998) Tetrachloroethylene. in: Occupational toxicants, Vol. 3, VCH, Weinheim

Henschler D, Lehnert G (Eds) (1994) Tetrachloroethene. in: Biological exposure values for occupational toxicants and carcinogens – Critical data evaluation for BAT and EKA values, Vol. 1, VCH, Weinheim, pp. 139–151

Jang J-Y, Kuang S-K, Chung HK (1993) Biological exposure indices of organic solvents of Korean workers. Int Arch Occup Environ Health 65, 219–222

Jang J-Y, Droz PO, Berode M (1997) Ethnic differences in biological monitoring of several organic solvents. I. Human exposure experiment. Int Arch Occup Environ Health 69, 343–349

Reitz RH, Gargas ML, Mendrala AL, Schumann AM (1996) *In vivo* and *in vitro* studies of perchloroethylene metabolism for physiologically based pharmacokinetic modeling in rats, mice and humans. Toxicol Appl Pharmacol 136: 289–306

Skender LJ, Karacic V, Prpic-Majic D (1991) A comparative study of human levels of trichloroethylene and tetrachloroethylene after occupational exposure. Arch Environ Health 46: 174–178

Völkel W, Friedewald M, Lederer E, Pahler J, Parker J, Dekant W (1998) Biotransformation of perchloroethene: dose-dependent excretion of trichloroacetic acid, dichloroacetic acid and N-acetyl-S-(trichlorovinyl)-L-cysteine in rats and humans after inhalation. Toxicol Appl Pharmacol 153: 20–27

Authors: K.H. Schaller, H.M. Bolt
Approved by the Working Group: 08.07.1997 and 20.02.2001

Trichloroethene, Addendum

EKA The correlations between external and internal exposure yield the following data:

Air Trichloroethene (ml/m^3)	(mg/m^3)	Urine Trichloroacetic acid (mg/l)
10	55	20
20	109	40
30	164	60
50	273	100

Sampling time: end of exposure or end of shift

Date of evaluation 2001

The toxicity of trichloroethene was originally reviewed in Volume 1 of this series (Henschler and Lehnert 1994). Developments in the meantime have made re-evaluation of the substance necessary. In 1996 trichloroethene was classified in Carcinogen category 1. This was the result of observations that kidney tumours are formed in man. Genotoxic mechanisms of action, together with the renal toxicity caused by exposure to high levels of the substance, play an important role. It is explained in a supplement to the MAK documentation that under the present circumstances a health-based MAK value cannot be set (Greim 2000).

1 Evaluation of the Biological Exposure Equivalents

1.1 Metabolism and kinetics

Like some other chlorinated hydrocarbons, trichloroethene is metabolized via two alternative metabolic pathways. The greatly predominating, oxidative, metabolic pathway leads via trichloroacetaldehyde (chloral) to trichloroethanol and trichloroacetic acid. Trichloroethanol (mainly in conjugated form) and trichloroacetic acid are the main metabolites excreted in the urine. The quantitative relationships were described in detail in the BAT documentation for trichloroethene (see the BAT documentation for trichloroethene from 1994 in this series).

A secondary pathway is the glutathione-dependent metabolic pathway, which, via glutathione and cysteine conjugates, leads to reactive metabolites which can be formed by the "beta-lyase" in kidney tissue (Goeptar *et al.* 1995).

The quantitative data documented in the literature for the metabolism of trichloroethene in man and species of experimental animal were used in the toxicokinetic models ("Physiologically Based Pharmacokinetics" (PBPK)) developed by several working groups (Clewell *et al.* 2000, Fisher *et al.* 1998, Poet *et al.* 2000). During the development of one of these models attention was drawn to the individual variability of oxidative metabolism in man, which manifests itself in the variable levels of trichloroethanol-glucuronide and trichloroacetic acid excreted with the urine and in the different importance of the metabolic pathway leading to dichloroacetic acid (Fisher *et al.* 1998). In this context, also marked quantitative differences in the oxidative metabolism of trichloroethene were described, which were determined *in vitro* with 23 human microsome preparations. This variability was associated with differences in the activity of the cytochrome P450 isoenzyme CYP2E1 (Lipscomb *et al.* 1997; see also Bolt *et al.* 2003).

For the quantitative relationship between external exposure to trichloroethene and the excretion of trichloroethanol and trichloroacetic acid, the reader is referred to Volume 1 of this series (Henschler and Lehnert 1994).

1.2 Critical toxicity

New insights have been gained into the nephrotoxicity and nephrocarcinogenicity of the substance in man. A comprehensive review of the findings is available (Brüning and Bolt 2000). The studies of the carcinogenic effects of trichloroethene are also described in detail in the MAK documentation (Greim 2000).

Further investigations of the underlying nephrotoxic mechanisms were possible on the basis of the cases of cancer of the kidneys after occupational exposure to trichloroethene described in two studies (Brüning *et al.* 2003, Henschler *et al.* 1995, Vamvakas *et al.* 1998).

An important starting point was the theory put forward on the basis of mechanistic considerations that renal cell tumours caused by exposure to trichloroethene occur in particular when the substance also caused toxic damage to the renal cells (Goeptar *et al.* 1995).

In one study, first of all the excretion of proteins was investigated in 17 patients with former occupational exposure to high levels of trichloroethene who developed renal cell carcinomas (Brüning *et al.* 1996). The proteins were separated from the urine using an SDS-PAGE (sodium dodecylsulfate-polyacrylamide gel electrophoresis) method (Bazzi *et al.* 1997). In these 17 patients tubular damage to the kidneys was found, which in six patients was described as severe, in six patients as moderate and in two patients as slight; in three patients it was described as of mixed tubular and glomerular origin. In comparison, among 35 patients with renal cell carcinomas who were not previously occupationally exposed to trichloroethene a normal excretion pattern was found in 18 patients. Tubular damage was described as severe in five patients, as moderate in three patients, and as slight in four patients. In four patients mixed glomerular damage was found. One type of glomerular damage was diagnosed in one patient. These investigations were continued after the collective was extended to include 39 renal cell carcinoma patients with a working history of high-level exposure to trichloroethene. A semi-quantitative classification in exposure groups was carried out in accordance with the procedure described in Vamvakas *et al.* (1998). The control group was made up of 50 kidney cancer patients and 100 healthy control persons, all without exposure to trichloroethene. Using the SDS-PAGE method, protein excretion patterns which indicated damage to the kidney tubule apparatus were found in 38 of 41 patients with cancer of the kidneys exposed to trichloroethene (93 %), but in only 23 of 50 patients with cancer of the kidneys not exposed to trichloroethene (46 %), and in 11 of the 100 healthy control persons. The results were further supported by data for the excretion of α_1-microglobulin (Brüning *et al.* 1999a, Bolt *et al.* 2004). The higher incidence of tubular damage in patients with cancer of the kidneys (even in those not exposed to trichloroethene) relative to in the healthy controls can be explained by the effects of nephrectomy; the effect of so-called hyperfiltration (following the loss of kidney proteins) is the most probable mechanism (Golka *et al.* 1997).

With the aim of investigating persistent nephrotoxic effects after exposure to trichloroethene independent of the formation of cancer of the kidneys, a retrospective study was carried out in 39 workers, who (between 1956 and 1975) were exposed to high concentrations of trichloroethene. The control group comprised 46 persons from the same factory, who were matched for age and were employed in the offices of the company; they were therefore not occupationally exposed to trichloroethene or other chemicals. As in the previous studies using the SDS-PAGE method, the results revealed an increased incidence of tubular damage to the kidneys as an effect of high-level exposure to trichloroethene. The exposed persons were found also to excrete increased amounts of glutathione *S*-transferase α in urine, which represents a specific marker for damage to the proximal kidney tubule. On the other hand, the amounts excreted of glutathione *S*-transferase π, a marker for damage to the distal tubule, were in the same range in the exposed persons and in the control group (Brüning *et al.* 1999b)

Thus, the theory derived from experiments with animals that the occurrence of renal cell carcinomas after exposure to high levels of trichloroethene also occurs on the grounds of tubular damage to renal cells (Goeptar *et al.* 1995) must also be considered for man. In addition, the different excretion of the isoenzymes of glutathione *S*-transferase (α and π) in urine in the observed cases indicates selective damage of the proximal tubule.

1.3 Selection of the indicators

1.3.1 Exposure

To date, the determination of trichloroethanol in blood and of the main metabolite trichloroacetic acid in urine have been used as indicators. The previous BAT value for trichloroethanol was directly based on the quantity of neurotoxic effects (avoidance of pre-narcotic symptoms) caused particularly by this metabolite. The BAT value for the excretion of trichloroacetic acid, however, was based on the correlation with the former MAK value of 50 ml trichloroethene per m^3 air. After the classification of trichloroethene in Carcinogen category 1, the primary reference to the neurotoxic effects is no longer valid. The established relationships between external exposure and the excretion of trichloroacetic acid can, however, still be used for drawing up an EKA correlation.

1.3.2 Effects

It can be said (see Section 1.2 "Critical toxicity") that cancer of the kidneys caused by trichloroethene occurs in particular when at the same time toxic effects on the proximal tubule apparatus of the kidneys can be detected. This is to be expected at high trichloroethene concentrations greatly above the former threshold limit in air of 50 ml/m^3.

On the other hand, subclinical damage to the kidney tubules is found also in about 15 % of the normal population (Brüning *et al.* 1999a, 1999b, Bolt *et al.* 2004); genotoxic effects of the trichloroethene metabolites ultimately effective in the kidneys are, however, probable. In the case of occupational contact with trichloroethene, special occupational-medical monitoring examinations are therefore recommended.

The determination of effect markers can be used in monitoring examinations; the excretion of specific proteins in urine can indicate as yet subclinical damage to the proximal tubules. The following determinations in urine are suitable:
– the protein pattern after electrophoretic separation using SDS-PAGE (as the screening method)
– α_1-microglobulin
– glutathione *S*-transferase α

α_1-Microglobulin seems to be the most suitable effect marker as it is stable in urine (unlike β_2-microglobulin also at low pH values) and is an easily determined microprotein.

The determination of the parameter α_1-microglobulin during the first examination serves as a screening procedure for tubular damage also of non-toxic genesis (excretion is increased e.g. also in the case of nephrosklerosis).

In addition, the glutathione *S*-transferase isoenzyme α in urine can be determined; increased excretion indicates specific damage to the proximal tubule and it therefore seems to be a sensitive marker for trichloroethene-induced nephrotoxicity (Brüning *et al.* 1999b). Also the glutathione *S*-transferase proteins seem to be sufficiently stable in urine. The prerequisite for an official recommendation is, however, the general availability of the method of investigation.

1.4 Methods

There is a recent study available of the analysis of derivatized trichloroacetic acid in the urine of exposed workers by capillary gas chromatography (O'Donnell *et al.* 1995). In addition, the reader is referred to the collection of methods "Analytische Methoden zur Prüfung gesundheitsschädlicher Arbeitsstoffe" (in English translation as "Analyses of Hazardous Substances in Biological Materials") (Angerer and Schaller 1985, 1991a, 1991b)

1.5 Exposure equivalents for carcinogenic substances (EKA)

In the BAT documentation for trichloroethene (in Volume 1 of this series; Henschler and Lehnert 1994) the relationships between external exposure and the excretion of trichloroacetic acid in urine were described in detail.

Taking into consideration the available studies in man, for the former MAK value for trichloroethene of 50 ml/m^3 this yielded a correlating BAT value of 100 mg trichloroacetic acid/l urine. On the basis of this data, the following EKA correlations can be established:

Air Trichloroethene		Urine Trichloroacetic acid
(ml/m^3)	(mg/m^3)	(mg/l)
10	55	20
20	109	40
30	164	60
50	273	100

Sampling time: end of exposure or end of shift.

The correlation is calculated assuming daily exposure (8 hours/day) at the level of the mean trichloroethene concentration in air given in the table.

1.6 Interpretation of data

The reader is referred to the G 14 guidelines for preventive medical examinations issued by the Employers' Liability Insurance Association (Berufsgenossenschaftliche Grundsätze für arbeitsmedizinische Vorsorgeuntersuchungen des Hauptverbandes der gewerblichen Berufsgenossenschaften (HVBG)). Here the intervals between follow-up examinations are also given.

2 References

Angerer J, Schaller KH (1985) Trichloroacetic acid (TCA) in urine. in: Henschler D (Ed.) Analyses of hazardous substances in biological materials, Vol. 1, VCH, Weinheim, pp. 209–217

Angerer J, Schaller KH (1991a) Trichloroethylene. in: Henschler D (Ed.) Analyses of hazardous substances in biological materials, Vol. 3, VCH, Weinheim, pp. 127–150

Angerer J, Schaller KH (1991b) Trichloroethene in blood. in: Henschler D (Ed.) Analyses of hazardous substances in biological materials, Vol. 3, VCH, Weinheim, pp. 127–150

Bazzi C, Petrini C, Rizza V, Arrigo G, Beltrame A, D'Amico G (1997) Characterization of proteinuria in primary glomerulonephritides. SDS-PAGE patterns: clinical significance and prognostic value of low molecular weight ("tubular") proteins. Am J Kidney Dis 29: 27–35

Bolt HM, Roos PH, Thier R (2003) The cytochrome P-450 isoenzyme CYP2E1 in the biological processing of industrial chemicals: consequences for occupational and environmental medicine. Int Arch Occup Environ Health 76: 174–185

Bolt HM, Lammert M, Selinski S, Brüning T (2004) Urinary alpha 1-microglobulin excretion as biomarker of renal toxicity in trichloroethylene-exposed persons. Int Arch Occup Environ Health 77: 186–190

Brüning T, Golka K, Makropoulos V, Bolt HM (1996) Preexistence of chronic tubular damage in cases of renal cell cancer after long and high exposure to trichloroethylene. Arch Toxicol 70: 259–260

Brüning T, Mann H, Melzer H, Sundberg AG, Bolt HM (1999a) Pathological excretion patterns of urinary proteins in renal cell cancer patients exposed to trichloroethylene. Occup Med 49: 299–306

Brüning T, Sundberg AG, Birner G, Lammert M, Bolt HM, Appelkvist EL, Nilsson R, Dallner G (1999b) Glutathione transferase alpha as a marker for tubular damage after trichloroethylene exposure. Arch Toxicol 73: 246–254

Brüning T, Bolt HM (2000) Renal toxicity and carcinogenicity of trichloroethylene: key results, mechanisms, and controversies. Crit Rev Toxicol 30: 253–285

Brüning T, Pesch B, Wiesenhutter B, Rabstein S, Lammert M, Baumüller A, Bolt HM (2003) Renal cell cancer risk and occupational exposure to trichloroethylene: results of a consecutive case-control study in Arnsberg, Germany. Am J Ind Med 43: 274–285

Clewell HJ, Gentry PR, Covington TR, Gearhart JM (2000) Development of a physiologically based pharmacokinetic model of trichloroethylene and its metabolites for use in risk assessment. Environ Health Perspect 108, Suppl 2: 283–305

Fisher JW, Mahle D, Abbas R (1998) A human physiologically based pharmacokinetic model for trichloroethylene and its metabolites, trichloroacetic acid and free trichloroethanol. Toxicol Appl Pharmacol 152: 339–359

Goeptar A, Commandeur JN, van Ommen B, van Bladeren PJ, Vermeulen NP (1995) Metabolism and kinetics of trichloroethylene in relation to toxicity and carcinogenicity: relevance of the mercapturic acid pathway. Chem Res Toxicol 8: 3–21

Golka K, Hartert-Raulf T, Schöps W, Kierfeld G, Brüning T, Bolt HM (1997) Subclinical changes in urinary protein excretion of renal cell patients: effect of high occupational exposure to trichloroethylene. Central Eur J Occup Environ Med 3: 167–174

Greim H (Ed.) (2000) Trichlorethen. Gesundheitsschädliche Arbeitsstoffe, Toxikologisch-arbeitsmedizinische Begründungen von MAK-Werten, 31. Lieferung, Wiley-VCH, Weinheim

Henschler D, Lehnert G (Eds) (1994) Trichloroethene. in: Biological exposure values for occupational toxicants and carcinogens – Critical data evaluation for BAT and EKA values, Vol. 1, VCH, Weinheim, pp. 171–183

Henschler D, Vamvakas S, Lammert M, Dekant W, Kraus B, Thomas B, Ulm K (1995) Increased incidence of renal cell tumors in a cohort of cardboard workers exposed to trichloroethene. Arch Toxicol 69: 291–299

Lipscomb JC, Garrett CM, Snawder JE (1997) Cytochrome P450-dependent metabolism of trichloroethylene: interindividual differences in humans. Toxicol Appl Pharmacol 142: 311–318

O'Donnell GE, Juska A, Geyer R, Faiz M, Stalder S (1995) Analysis of trichloroacetic acid in the urine of workers occupationally exposed to trichloroethylene by capillary gas chromatography. J Chromatogr A 709: 313–317

Poet TS, Corley RA, Thrall KD, Edwards JA, Tanojo H, Weitz KK, Hui X, Maibach HI, Wester RC (2000) Assessment of the percutaneous absorption of trichloroethylene in rats and humans using MS/MS real-time breath analysis and physiologically based pharmacokinetic modeling. Toxicol Sci 56: 61–72

Vamvakas S, Brüning T, Thomasson B, Lammert M, Baumüller A, Bolt HM, Dekant W, Birner G, Henschler D, Ulm K (1998) Renal cell cancer correlated with occupational exposure to trichloroethene. J Cancer Res Clin Oncol 124: 374–382

Author: H.M. Bolt
Approved by the Working Group: 20.02.2001

Documentation for Carcinogenic Substances
Without Biological Exposure Equivalents

Antimony and its inorganic compounds

Antimony

BAT	not established
Date of evaluation	2002
Synonyms	Stibium
Formula	Sb
CAS No.	7440-36-0
Molecular weight	121.75
Melting point	630.5°C
Boiling point	1635°C
Vapour pressure at 20°C	1420 mm Hg
MAK [last established: 2005]	Carcinogen category 2

Antimony trioxide

EKA	not established
Date of evaluation	2002
Synonyms	Antimonous oxide Antimony(III) oxide White antimony Antimony white Anhydrous stibious acid Antimonous acid anhydride naturally occurring as: Senarmontite (cubic) Valentinite (rhombic) Flowers of antimony and antimony bloom
Chemical name (CAS)	Antimony(III) oxide

CAS No.	1309-64-4, 1327-33-9
Formula	Sb_2O_3
Molecular weight	291.52
Melting point	656°C
Boiling point	1500°C
MAK [last established: 2005]	Carcinogen category 2

Antimony hydride

BAT	not established
Date of evaluation	2002
Synonyms	Antimony(III) hydride Stibine Stibane Monostibane
Chemical name (CAS)	Antimony(III) hydride
CAS No.	7803-52-3
Formula	SbH_3
Molecular weight	124.77
Melting point	–88°C
Boiling point	–17.1°C
MAK [last established: 1958]	0.52 mg/m^3 ≙ 0.1 ml/m^3 (ppm)

The most important industrial use of antimony is based on its ability to positively influence alloys, e.g. to make lead harder. In addition, antimony compounds are used for cosmetic and therapeutic purposes. In recent times this metal has increased in importance as a flame retardant and in the production of semi-conductors. The most important uses of antimony and its compounds can be seen in Table 1 (Plunkert 1981).

Table 1. Uses for antimony compounds

Used in	Antimony compound	Product/function
industry	antimony (Sb)	semi-conductors
	antimonial lead (PbSb)	accumulator grids, projectiles and pellets
	antimonial tin (SbZn)	bearing metal
	antimony trichloride ($SbCl_3$)	catalyst for organic synthesis, tanning auxiliary, burnishing agent for gunmetal
	antimony trioxide (Sb_2O_3)	opacifier in white enamels, flame retardant for plastics (e.g. PVC), textiles, paints and paper
	antimony trisulfide (Sb_2S_3)	vulcanization assistant
	antimony pentasulfide (Sb_2S_5)	matches, fire-fighting equipment, pigment for glass
medicine	trivalent and pentavalent organic antimony compounds	therapeutic agent for tropical parasitosis
	technetium-99 m antimony sulfide	tracer in experimental prostate tumour diagnosis
agriculture	potassium antimonyl tartrate	ant extermination agent

1 Pharmacokinetics

1.1 Absorption

At the workplace the inhalation of dusts containing antimony or gaseous antimony hydride are the most important sources of exposure. Ingestion and dermal absorption are of lesser importance with workplace exposure.

As a result of its poor solubility, antimony trioxide is absorbed only slowly into body fluids via the mucous membranes of the respiratory and gastrointestinal tracts. The pulmonary absorption coefficients were determined from the ratio of antimony in the air to antimony in the blood of employees from a battery factory (Kentner et al. 1993). In casters exposed to antimony trioxide dusts the absorption coefficient was 1.74; for tank formers, who were exposed to both antimony hydride and antimony trioxide, 1.22 (Kentner et al. 1995).

1.2 Metabolism and kinetics

Inorganic antimony, unlike arsenic, is not methylated *in vivo* (Bailly et al. 1991, Buchet and Lauwerys 1985).

Absorbed trivalent antimony is rapidly eliminated from the blood plasma and absorbed mainly by the lungs, liver and kidneys. From these organs it is slowly released into the blood again and the trivalent antimony compounds are excreted mainly via the kidneys. A small amount of the antimony is eliminated via the intestine after binding in the liver with glutathione (Bailly *et al.* 1991, Gerhardsson *et al.* 1982).

1.3 Elimination

There are data available for the excretion kinetics of antimony from both animal experiments and observations in man. In an experiment with hamsters given intratracheal doses of antimony trioxide, two phases were observed during elimination of the substance. In the initial phase the elimination half-time from lung tissue was about 40 hours, in the second phase 20 to 40 days (Leffler *et al.* 1984).

In investigations of employees from a battery factory a rapid renal elimination phase with a half-time of about 34 hours and a slow phase with a half-time of about 90 hours were determined (Kentner *et al.* 1993).

Casters with exposure to antimony trioxide and tank formers who absorbed antimony hydride were found to have renal half-lives of about 95 hours (Kentner *et al.* 1995). The elimination coefficient (antimony in blood to antimony in urine) was 0.68.

Comparable half-lives were calculated for persons occupationally exposed to pentavalent antimony compounds (Bailly *et al.* 1991). Other studies suggest that renal elimination of pentavalent inorganic antimony compounds is more rapid (overview in Winship 1987).

Further data for biological half-times from animal experiments can be found in Apostoli *et al.* (1994), Elinder and Friberg (1986), Norseth and Martinsen (1988) and Winship (1987).

2 Critical Toxicity

The toxicity of antimony and its compounds has been discussed in detail in several publications (Elinder and Friberg 1986, Kentner and Leinemann 1994, Norseth and Martinsen 1988, Winship 1987).

Only few cases of acute or chronic intoxication with antimony, antimony trioxide or antimony hydride are known to date in industry.

The symptoms of chronic intoxication with antimony, antimony trioxide and antimony hydride have been described (Kentner and Leinemann 1994):

Compound	Chronic intoxication
antimony (Sb)	dermatitis, myocardial and liver damage, pneumoconiosis, eosinophilia, increased spontaneous abortions
antimony trioxide (Sb_2O_3)	irritation of the mucous membrane, coughing, nausea, vomiting, aching joints, stomach ache, muscle pains, loss of appetite, myocardial damage, dizziness, sleeplessness, dermatitis and conjunctivitis
antimony hydride (SbH_3)	there are no data available for chronic intoxication, possible symptom: haemoglobinuria

For the carcinogenic effects of antimony trioxide the reader is referred to the MAK documentation from 1983 classifying antimony trioxide in the form of inhalable dusts in Carcinogen category 2 (Henschler 1983). The documentation setting the TLV(Threshold Limit Value)-TWA(Time-Weighted Average) of 0.5 mg/m^3 for antimony and its compounds, and classifying antimony trioxide from production in Category A2 are from 1991 (ACGIH 1996).

3 Exposure and Effects

No occupational-medical studies of the exposure–response relationship have been published. For the relationship between external and internal exposure two field studies are available.

Kentner et al. (1993, 1995) investigated 21 employees from a battery factory. Seven worked as casters and 14 as tank formers. The casters were exposed to antimony trioxide, the tank formers to antimony trioxide and antimony hydride. The external exposure was determined by means of personal air sampling. Both antimony trioxide dust and gaseous antimony hydride were analysed. To quantify the internal exposure, blood and urine samples were collected at the end of the shift and at the end of the working week.

The medians and the total antimony concentration ranges in air, urine and blood were as follows:

	Casters (Sb_2O_3)			Tank formers (Sb_2O_3 and SbH_3)		
	median	min.	max.	median	min.	max.
air (mg Sb/m^3)	0.0045	0.0012	0.0066	0.0124	0.0006	0.0415
urine (µg Sb/g creatinine)	3.9	2.8	5.6	15.2	3.5	23.4
blood (µg Sb/l blood)	2.6	0.5	3.4	10.1	0.5	17.9

The external exposure in the two groups investigated was lower by a factor of 10 than the current MAK and TRK values. The exposure of the tank formers was three to four times higher than that of the casters. Evaluation of the data from investigations revealed that the absorption and elimination properties of antimony oxide and antimony hydride are comparable. The two collectives were therefore grouped together for the correlation analysis. The correlation coefficient for the relationship between the external exposure and the excretion of antimony with the urine, expressed as creatinine, was 0.75 ($y = 0.52x + 5.87$) and for the relationship between the external exposure and antimony concentration in blood 0.8 ($y = 0.44x + 2.38$). There were 21 pairs of values.

The correlation between external exposure and the renal excretion of antimony was calculated also in 20 workers exposed to antimony pentoxide and sodium antimonate in a non-metal smelting plant (Bailly et al. 1991). Personal air sampling was carried out to evaluate the total amount of antimony in the air. A urine sample was collected at the beginning and the end of the working day to determine the antimony and creatinine. The investigation took place in the second half of the working week. The first series of analyses ($N = 26$) was carried out while the workers handled solutions containing antimony or wet pastes (wet process). In a second series (N = 14) the analyses took place during the grinding, sieving and packing of the antimony products (dry process). As expected, different levels of exposure were found for activities during the wet process and dry process.

The mean values and standard deviations for the exposure values are shown below:

Workplace	Number of analyses	Air (μg total Sb/m^3)	Urine (μg Sb/g creatinine)	
			pre-shift	post-shift
wet process	26	86 ± 78	8.2 ± 3.9	12.3 ± 5.0
dry process	14	927 ± 985	58.4 ± 62.5	110 ± 76

The correlation between external and internal exposure (r = 0.83) was found to be significant. There is an even closer correlation if the increase in the antimony concentrations during the shift is used as the correlate (r = 0.86, N = 35). The authors calculated for an antimony concentration in air of 0.5 mg/m^3 an increase in the renal excretion of antimony during the shift of 35 µg antimony per g creatinine. Determination of the antimony concentration in urine is suggested as the parameter for evaluating the intensity of current exposure.

4 Selection of the Indicators

Important biological parameters for determining the exposure to antimony are
– the antimony concentration in whole blood
– the antimony excreted in the urine

The determination of antimony in whole blood is time-consuming. For this reason and because of the longer elimination kinetics, the determination of antimony in urine is preferable (Gebel et al. 1998).

The two field studies showed that the creatinine-related values are more suitable for the biological monitoring of excretion than the litre values. As a result of the biological half-time of about 90 hours, antimony accumulates slightly in urine over the working week (Kentner et al. 1993). The suggested sampling time is, therefore, post-shift at the end of the working week.

5 Methods

In the series "Analyses of Hazardous Substances in Biological Materials" a reliable and tested method for the determination of antimony in urine is described (Angerer and Schaller 1994). The antimony in urine is determined by means of electrothermal atomic absorption spectrometry without previous preparation of the sample. Triton X-100 and a matrix modifier containing palladium are added to the urine sample (Fleischer and Schaller 1987).

Hydride atomic absorption spectrometry can also be used, however, to determine the antimony in urine. After the addition of sodium iodide and hydroxylamine, the urine sample is reduced with sodium borohydride. The stibine formed is thermally decomposed and determined by means of atomic absorption spectrometry (Angerer and Schaller 1988). The analytical detection limit of the two atomic absorption spectrometric procedures is about 1 µg antimony per litre urine.

6 Background Exposure

The amount of antimony excreted in persons not occupationally exposed to the substance is very small. With a detection limit for antimony in urine of 0.6 µg/l, in persons not occupationally exposed no antimony could be detected (Fleischer and Schaller 1987). Sources of background exposure in the environmental-medical range are not known.

7 Evaluation of BAT Values and EKA Correlations

To evaluate threshold limit values in biological materials or EKA correlations, the principal parameter available is the determination of the antimony excreted in urine. There are two field studies in which data for the external exposure to antimony and the antimony excreted in urine are described.

In the field study carried out by Kentner et al. (1995) with exposure to antimony trioxide dusts and gaseous antimony hydride, a significant relationship ($r = 0.75$, $y = 0.52x + 5.87$) between the external exposure and the creatinine-related excretion of antimony in urine was calculated. On the basis of this correlation the authors suggested the following threshold limit values in biological materials, taking into account the MAK and former TRK values:

Current threshold limit value in air	Antimony excreted in urine (µg antimony/g creatinine)
0.1 mg antimony trioxide/m^3 (former TRK)	60
0.3 mg antimony trioxide/m^3 (former TRK)	160
0.5 mg antimony hydride/m^3 (MAK)	260

With these suggested threshold limit values in biological materials it must be noted in particular that in the production of starter batteries the antimony concentrations in air are generally below 0.05 mg/m^3. Extrapolation to values of 0.3 and 0.5 mg/m^3 is, as a matter of course, subject to considerable error.

In the study of Bailly et al. (1991) the external and internal exposure of 20 workers exposed to antimony pentoxide and sodium antimonate were investigated. A total external exposure to antimony of 0.5 mg/m^3 corresponded to an increase in the renal excretion of antimony during the shift of 35 µg/g creatinine. If the antimony concentration in the pre-shift sample is assumed to be very low, this would correspond to a threshold limit value for the post-shift urine sample of 35 µg/g creatinine with an antimony concentration in air of 0.5 mg/m^3.

At present the following reasons, however, speak against setting BAT values or EKA correlations for the occupational-medical evaluation of occupational exposure to antimony:

1. There are insufficient data available. Only two field studies of external and internal exposure to the substance have been published.
2. In both field studies different antimony species (antimony trioxide, antimony hydride and antimony(V) compounds) occur at the workplaces.
3. The two suggestions for threshold limit values in biological materials use different parameters. In one case an increase in the antimony concentration in urine over the workshift is given, in the other case the parameter used is the creatinine-related excretion of antimony in urine samples collected post-shift at the end of the working week.
4. The suggested threshold limit values in the two field studies differ considerably. Kentner et al. (1995) suggest for an external exposure of 0.5 mg total antimony per m^3 a value of 260 µg/g creatinine for the antimony excreted, while the study of Bailly et al. (1991) yields a value of 35 µg/g creatinine at this level of external exposure. This comparison is limited by the fact that the two working groups chose different reference points for the threshold limit value suggested (the antimony eliminated in urine is expressed as the increase during the shift and in the post-shift sample).
5. This discrepancy and the uncertainty whether one threshold limit value or correlation can be evaluated for the different antimony species, at present do not allow the evaluation of scientifically grounded threshold limit values in biological materials.

8 References

ACGIH (American Conference of Governmental Industrial Hygienists) (1996) Antimony and compounds. in: Documentation of the threshold limit values and biological exposure indices, 6th edition, ACGIH, Cincinnati, Ohio, 73–75

Angerer J, Schaller KH (1988) Antimony. in: Henschler D (Ed.) Analyses of hazardous substances in biological materials, Vol. 2, VCH, Weinheim

Angerer J, Schaller KH (1994) Antimony. in: Greim H (Ed.) Analyses of hazardous substances in biological materials, Vol. 4, VCH, Weinheim

Apostoli P, Porru S, Alessio L (1994) Biological indicators for the assessment of human exposure to industrial chemicals. Antimony. Report EUR 14815 EN, European Commission, Brussels, 5–21

Bailly R, Lauwerys R, Buchet JP, Mahieu P, Konings J (1991) Experimental and human studies on antimony metabolism: their relevance for the biological monitoring of workers exposed to inorganic antimony. Brit J Ind Med 48: 93–97

Buchet JP, Lauwerys R (1985) Study of inorganic arsenic methylation by rat liver *in vitro*: relevance for the interpretation of observations in man. Arch Toxicol 57: 125–129

Elinder CG, Friberg L (1986) Antimony. in: Friberg L, Nordberg GF, Vouk VB (Ed.) Handbook on the toxicology of metals, 2nd edition, Elsevier, Amsterdam-New York-Oxford, 26–42

Fleischer M, Schaller KH (1987) Direkte quantitative Bestimmung von Arsen und Antimon in Körperflüssigkeiten mit dem Graphitrohrofen. in: Welz B (Ed.) 4. Colloquium Atomspektrometrische Spurenanalytik, Bodenseewerk Perkin-Elmer & Co GmbH, Überlingen, 577–586

Gebel T, Claussen K, Dunkelberg H (1998) Human biomonitoring of antimony. Int Arch Occup Environ Health 71: 221–224

Gerhardsson L, Brune D, Nordberg GF, Wester PO (1982) Antimony in lung, liver and kidney tissue from deceased smelter workers. Scand J Work Environ Health 8: 201–208

Henschler D (Ed.) (1983) Antimontrioxid (Atembare Stäube). Gesundheitsschädliche Arbeitsstoffe, Toxikologisch-arbeitsmedizinische Begründungen von MAK-Werten, 9. Lieferung, VCH, Weinheim

Kentner M, Leinemann M, Schaller KH, Weltle D (1993) Antimonbelastungen in der Bleiakkumulatoren-Herstellung. in: Triebig G, Stelzer O (Eds) Dokumentationsband über die Verhandlungen der Deutschen Gesellschaft für Arbeitsmedizin und Umweltmedizin e.V., Gentner-Verlag, Stuttgart, 111–116

Kentner M, Leinemann M (1994) Umwelt- und arbeitsmedizinische Bedeutung von Antimon und seinen wichtigsten Verbindungen. Zbl Arbeitsmed 44: 46–55

Kentner M, Leinemann M, Schaller KH, Weltle D, Lehnert G (1995) External and internal antimony exposure in starter battery production. Int Arch Occup Environ Health 67: 119–123

Leffler G, Gerhardsson L, Brune D, Nordberg GF (1984) Lung retention of antimony and arsenic in hamsters after the intratracheal instillation of industrial dust. Scand J Work Environ Health 10: 245–251

Norseth T, Martinsen I (1988) Biological monitoring of antimony. in: Clarkson TW, Friberg L, Nordberg GF, Sager PR (Eds) Biological monitoring of toxic metals, Plenum Press, New York, 337–367

Plunkert PA (1981) Antimony. Minerals Yearbook 1, US Department of the Interior, US Geological Survey, Washington DC, 93–102

Winship KA (1987) Toxicity of antimony and its compounds. Adv Drug React Ac Pois Rev 2: 67–90

Author: K.H. Schaller
Approved by the Working Group: 11.01.2002

Beryllium and its inorganic compounds

EKA	not established
Date of evaluation	2002
Formula	Be
CAS No.	7440-41-7
Molecular weight	9.01
Melting point	1283°C
Boiling point	2970°C
MAK [last established: 2003]	Carcinogen category 1

Beryllium is the metal with the lowest atomic weight. In the earth's crust, its compounds are widespread in low concentrations. Beryllium has been processed industrially since the 1930s. Advantageous are its light weight and high durability, even at higher temperatures.

It was recognised early on that beryllium and its compounds cause damage to the health. The acute and chronic lung diseases caused by the inhalation of dust and vapour were observed first in the 1940s and led to the inclusion of beryllium and its compounds in the German list of occupational diseases (BK No. 1110). In the subsequent years diseases caused by beryllium were observed only sporadically. Not until the 1990s, and particularly in the USA, did beryllium-related lung diseases occur once again. This led to the American Conference of Governmental Industrial Hygienists (ACGIH) suggesting that the threshold concentration in air be reduced from 2 µg/m^3 to 0.2 µg/m^3 (Ratney 2001). Beryllium and its compounds were classified as carcinogenic in man both by the ACGIH and the International Agency for Research on Cancer (IARC). In 2003 beryllium and its inorganic compounds were classified as carcinogenic in man also by the DFG Commission for the Investigation of Health Hazards of Chemical Compounds in the Work Area (Greim 2003). Recent investigations have confirmed that beryllium has carcinogenic effects on the lung in man (Sanderson *et al.* 2001).

1 Metabolism and Kinetics

1.1 Absorption

The main route of absorption in occupationally exposed persons is inhalation. There are no data in man for the extent to which inhaled beryllium is deposited in the lungs or absorbed. The particle size and solubility are assumed to be the decisive factors which influence the extent of deposition and clearance in the lungs. Animal experiments have shown that beryllium is deposited in the lungs and absorbed only slowly. The pulmonary clearance of inhaled beryllium is two-phase with a rapid elimination phase during the first one to two weeks after the end of exposure and a second, slower elimination phase.

Experiments with guinea pigs showed that after exposure to beryllium concentrations of 10 mg/m^3 for 8 and 16 hours, elimination with the urine of inhaled beryllium begins after two hours. The maximum amount eliminated is reached 10 to 30 hours after the end of exposure (Stiefel *et al.* 1980).

Absorption of beryllium through the skin contributes only little to the body burden of occupationally exposed persons. The absorption of ingested beryllium is low (WHO 1990).

1.2 Toxicokinetics

In blood samples from animals and man, 60 % to 70 % of the beryllium was shown to be bound to pre-albumins and gamma-globulins. Absorbed beryllium is transported to all tissues. Analytical investigations after the administration of beryllium showed that in most organs measurable beryllium concentrations could be detected.

Beryllium is eliminated also from the blood in a rapid and a slow phase. The biological half-time for the rapid phase is in the range of 1 to 60 days (Rhoads and Sanders 1985, Sanders *et al.* 1975). The slow elimination phase has a half-time of 0.6 to 2.3 years.

1.3 Elimination

The elimination of beryllium takes place primarily via the urine; it is excreted with the faeces only in small amounts (Stiefel *et al.* 1980). Colloidal beryllium bound in plasma cannot diffuse through the glomerular membranes, but is excreted via the tubules (Reeves 1986).

2 Critical Toxicity

Intoxication with beryllium and its compounds leads to three different types of symptoms, as described below.

After the inhalation of beryllium compounds, in particular soluble beryllium salts in relatively high concentrations, in earlier case reports acute effects on the respiratory passages were observed in the form of rhinitis, pharyngitis, tracheobronchitis, pneumonia and pulmonary oedema (Henschler 1992).

The chronic lung disease known as berylliosis is the result of the inhalation of mainly poorly soluble beryllium compounds, e.g. beryllium oxide, for many years. Berylliosis is a type of pulmonary granulomatosis, in which a cell-mediated immune reaction is of pathogenetic importance (Henschler 1992, Cullen *et al.* 1986).

Beryllium and beryllium compounds are carcinogenic in man. Several epidemiological studies, in particular the recently published results of an American cohort study with an embedded case–control study, confirm there is significantly increased mortality from lung cancer after long-term exposure (Sanderson *et al.* 2001).

A detailed description of the diseases caused by beryllium and its compounds can be found in the MAK documentation (Greim 2005) and in the ACGIH documentation "Beryllium and Compounds" (ACGIH 2001, 2003).

3 Exposure and Effects

3.1 Relationship between external and internal exposure

There are only few data available from field studies. Results from laboratory experiments or modelling studies are not available.

One study investigated the renal excretion of beryllium in employees of a dental laboratory (Apostoli *et al.* 1989b). The beryllium concentrations were in the range from 0.05 to 1.7 µg/l urine (mean value 0.34 µg/l). The amounts of beryllium renally excreted by the employees were significantly higher than those of persons not exposed. A close, statistically significant relationship between the renal excretion of beryllium and the duration of exposure was not found.

Another study investigated the beryllium concentrations in urine samples from 14 workers who cleaned oil-fired boilers (Cammarano *et al.* 1985). The mean beryllium level was 1.1 µg/l urine before the shift and 1.8 µg/l urine after the end of the shift.

External exposure to beryllium concentrations of 2 µg/m^3 air correlated with renally excreted beryllium in concentrations of 7 µg/l urine (Zorn *et al.* 1988). From this relationship, for a beryllium concentration in air of 0.2 µg/m^3 the amount excreted was calculated to be 0.7 µg/l urine. The validity of this correlation is, however, highly questionable.

There is some doubt about the reliability of the analytical methods used for the studies named, as explained in Section 5 "Methods". The published beryllium concentrations are much higher than one would expect today. Evidently both the sensitivity and the specificity of the analytical methods used earlier were very limited.

A cross-sectional study investigated 27 gemstone cutters who worked in 12 different companies for a mean period of 24.8 ± 15.4 years (Wegner et al. 2000). The 27 gemstone cutters were employed in the processing of gems containing beryllium for a mean period of 21 ± 13 hours/week. Beryllium was detected in 17 pre-shift urine samples and in 12 post-shift urine samples. The median pre-shift value was 0.09 µg/l, the range was between < 0.06 and 0.56 µg/l urine. The corresponding values for the post-shift sample were in the range from < 0.06 to 0.29 µg/l urine. At the workplace with the highest level of external exposure to beryllium, beryllium was detected in all persons (n = 9; mean value 0.18 ± 0.19 µg/l in the pre-shift samples and 0.12 ± 0.15 µg/l in the post-shift samples). A correlation analysis between external and internal exposure was not described.

In another group of 30 gemstone cutters, who processed gemstones containing beryllium for only 0.7 ± 1.2 hours/week, unlike the group in the study mentioned above, beryllium could not be detected in any of the urine samples.

Another study investigated 65 workers employed in two electric melting furnaces and two foundries for the production of copper alloys (Apostoli and Schaller 2001). Static sampling was carried out to determine the external exposure, post-shift urine samples were collected to quantify the internal exposure. The results of the ambient and biological monitoring are summarized below.

Table 1. Results of ambient and biological monitoring (Apostoli and Schaller 2001)

Groups	Beryllium concentration in air (µg/m^3)			Beryllium concentration in urine (µg/l)		
	N	Median	Range	N	Median	Range
Electric steel production						
Furnace	14	0.12	0.03–0.18	25	0.07	0.02–0.45
Casting	10	0.03	0.02–0.05	18	0.05	0.02–0.40
Foundry						
Furnace	7	0.27	0.04–0.8	12	0.25	< 0.02–0.49
Casting	8	0.31	0.1–0.9	10	0.15	< 0.02–0.54
Controls	–	–		30	n.d.	

n.d. = not detected (detection limit = 0.02 µg/l)

A significant correlation was calculated for the relationship between external and internal beryllium exposure. A beryllium concentration of 0.2 µg/m^3 corresponded to the renal excretion of beryllium in concentrations of about 0.15 µg/l.

3.2 Relationship between internal exposure and effects

Investigations of the relationship between the internal exposure and effects are not known.

4 Selection of the Indicators

According to the current state of the art, the determination of beryllium in urine is the most suitable method of evaluating internal exposure to beryllium (Apostoli et al. 1989a). The determination of beryllium in blood has not acquired any importance. Only few data have been collected (Reeves 1986). This also applies for the investigation of nasal secretion (Eakins and Lally 1973).

Exposure to beryllium and its compounds can lead to hypersensitization (Greim 2002, Newman et al. 1996). This can be detected by the so-called beryllium lymphocyte proliferation test (BLPT) in bronchoalveolar lavage fluid or in blood. The BLPT is based on the fact that the lymphocytes sensitized by beryllium react to produce proliferation. The BLPT seems to be the earliest way of identifying chronic beryllium disease. Although persons with positive results in the BLPT do not yet show any clinical symptoms, these persons could be diagnosed as cases of chronic beryllium disease. The BLPT is subject to high variability both from laboratory to laboratory and also within the laboratory (Deubner et al. 2001). The diagnostic significance of the test is therefore questionable and the sensitivity and specificity of the test are still uncertain. The BLPT does, however, have a predictive value in the investigation of persons in whom there is evidence of chronic beryllium disease. The use of the BLPT for screening has not yet proved to be practicable and reliable. At present, the BLPT cannot be recommended as an effect parameter.

5 Methods

Beryllium in urine can be determined by means of graphite furnace atomic absorption spectrometry (GF-AAS), inductively coupled plasma spectroscopy (ICP-OES) and inductively coupled plasma mass spectrometry (ICP-MS). Analytically reliable methods have been published by the working group in "Analyses of Hazardous Substances in Biological Materials" (Angerer and Schaller 1997). The detection limit is 0.1 µg/l urine. Modern developments in analytical techniques, in particular with the use of ICP-MS, have improved the detection limits to levels as low as 0.02 to 0.06 µg/l (Apostoli and Schaller 2001, Schramel et al. 1997, Wegner et al. 2000).

In view of this, the results of earlier publications must be evaluated critically; both the sensitivity and the specificity of the methods used earlier were inadequate for reliably determining low beryllium concentrations in urine. This is true, in particular, for earlier publications of beryllium concentrations in urine samples from the general population (see Section 6 "Background Exposure").

In accordance with present-day standards, ICP-MS or optimized GF-AAS methods should be used for the determination of beryllium in urine.

The urine samples must be collected in plastic vessels at the end of the shift and at the end of the working week. The danger of contamination is relatively small as only very low beryllium concentrations are present in the environment. At the workplace contamination can occur. The urine samples should therefore be given after showering and changing into street clothes.

6 Background Exposure

The general population takes up beryllium with food, drinking water and air. In general the daily uptake of beryllium is very low. Smoking might also be an additional source of exposure for the general population (WHO 1990).

The results of investigations with modern analytical methods show that the beryllium concentration in urine of persons not occupationally exposed is considerably below 1 µg/l. Table 2 shows the beryllium concentrations in the urine of persons not occupationally exposed to the substance.

Table 2. Beryllium concentrations in the urine of persons not occupationally exposed

References	Number of persons	Method	Beryllium concentrations in urine $\overline{x}$	SD
Minoia et al. 1985	56	GF-AAS	0.6	0.2 µg/l
Minoia et al. 1990	579	GF-AAS	0.4	range < 0.02–0.82 µg/l
Apostoli et al. 1989b, 1992	163	GF-AAS	0.2	0.16 µg/l range < 0.03–0.8 µg/l
Paschal et al. 1998	496	ICP-MS	median: 0.61 µg/l or 0.13 µg/g creatinine 95th percentile: 0.8 µg/l or 0.62 µg/g creatinine	
Wegner et al. 2000	34	GF-AAS	< 0.06 µg/l	
Apostoli and Schaller 2001	30	ICP-MS	< 0.02 µg/l	

To summarize, in the urine samples of persons not occupationally exposed to the substance, beryllium could not be detected in current studies with the ICP-MS technique and optimized GF-AAS techniques.

7 Evaluation of Exposure Equivalents

1. The current database is insufficient for establishing an EKA correlation for the parameter beryllium in urine. In earlier studies methods were used for urinalysis which were not of adequate sensitivity and specificity for producing reliable analytical results. In most cases adequate quality control was not carried out. The analytical developments make clear that in future studies modern analytical techniques, such as ICP-MS etc., should be used. In view of this, at present only two studies can be used for the possible evaluation of exposure equivalents.
2. The amounts of beryllium excreted in the urine of persons not occupationally exposed to the substance are below the detection limits of modern analytical procedures (e.g. less than 0.02 µg/l urine).
3. Recent studies have shown that the parameter beryllium in urine can evidently be used as an indicator of short-term exposure to beryllium.
4. The TLV-TWA for beryllium and its compounds suggested by the ACGIH is 0.0002 mg/m^3. With exposure at the level of the TLV value and an assumed volume of air of 10 m^3 per shift and 100 % absorption, the amount of beryllium absorbed via inhalation would be 2 µg per day. Most of the absorbed beryllium is eliminated with the urine. In occupationally exposed persons the amount of beryllium excreted in the urine should therefore be above that of persons not exposed. Extrapolation to yield the current TRK values of 0.002 and 0.005 mg/m^3 is not possible.

8 References

ACGIH (American Conference of Governmental Industrial Hygienists) (2001) Beryllium and compounds. TLV® Chemical substances 7th edition, Documentation, ACGIH® Publication #7DOC-059. ACGIH®, Cincinnati

ACGIH (American Conference of Governmental Industrial Hygienists) (2003) Beryllium and compounds. TLV® Chemical substances draft, Documentation, Notice of intended change, ACGIH® Publication #7NIC-042. ACGIH®, Cincinnati

Angerer J, Schaller KH (1997) Beryllium – Standard addition procedure. in: Greim H (Ed.) Analyses of hazardous substances in biological materials, Vol. 5, VCH, Weinheim

Apostoli P, Schaller KH (2001) Urinary beryllium – a suitable tool for assessing occupational and environmental beryllium exposure? Int Arch Occup Environ Health 74: 162–166

Apostoli P, Porru S, Minoia C, Alessio L (1989a) Biological indicators for the assessment of human exposure to industrial chemicals: beryllium, carbon monoxide etc. Industrial Health and Safety, Commission of the European Communities, Brussels–Luxemburg, 5–21

Apostoli P, Porru S, Alessio L (1989b) Behaviour of urinary beryllium in general population and in subjects with low-level occupational exposure. Med Lav 80: 390–396

Apostoli P, Minoia C, Gilberti ME, Ronchi A (1992) Determination of beryllium in urine by Zeeman GFAAS. in: Minoia C, Caroli S (Eds) Applications of Zeeman graphite furnace atomic absorption spectrometry in the chemical laboratory and in toxicology, Pergamon Press, Oxford–New York–Seoul–Tokyo, 495–516

Cammarano G, Catenacci G, Minoia C (1985) Esposizione a metalli in addetti alla pulitura di caldaie ad olio combustibile in una centrale termoelettrica. in: Atti 48° Congresso Nazionale della Società Italiana di Medicina del Lavoro e di Igiene Industriale, Pavia 18–21 Settembre 1985. Monduzzi Ed., Bologna, 179–183

Cullen MR, Cherniack MG, Kominsky JR (1986) Chronic beryllium disease in the United States. Semin Respir Med 7: 203–209

Deubner DC, Goodman M, Iannuzzi J (2001) Variability, predictive value, and uses of the beryllium blood lymphocyte proliferation test (BLPT): preliminary analysis of the ongoing workforce survey. Appl Occup Environ Hyg 16: 521–526

Eakins JD, Lally AE (1973) The determination of beryllium in nose blow and faecal samples using a (gamma, n) monitor. Ann Occup Hyg 16: 33–39

Greim H (Ed.) (2005) Beryllium and its compounds. in: Occupational toxicants, Vol. 21, Wiley-VCH, Weinheim

Henschler D (Ed.) (1992) Beryllium and its compounds. in: Occupational toxicants, Vol. 3, VCH, Weinheim

Minoia C, Pozzoli L, Cavalleri A, Capodaglio E (1985) Definizione di valori di riferimento di 30 elementi in trace nei liquidi biologici. in: Atti del 48° Congresso Nazionale Società Italiana Medicina del Lavoro e di Igiene Industriale, Pavia, 18–21 Settembre 1985, Monduzzi Ed. Bologna, 317–323

Minoia C, Sabbioni E, Apostoli P, Pietra R, Pozzoli L, Gallorini M, Nicolaou G, Alessio L, Capodaglio E (1990) Trace element reference values in tissues from inhabitants of the European Community. I. A study of 46 elements in urine, blood and serum of Italian subjects. Sci Total Environ 95: 89–105

Newman LS, Lloyd J, Daniloff E (1996) The natural history of beryllium sensitization and chronic beryllium disease. Environ Health Perspect 104S: 937–943

Paschal DC, Ting BG, Morrow JC, Pirkle JL, Jackson RJ, Sampson EJ, Miller DT, Caldwell KL (1998) Trace metals in urine of United States residents: reference range concentrations. Environ Res 76: 53–59

Ratney RS (2001) Is beryllium disease a fossil? – Not yet. Int Arch Occup Environ Health 74: 159–161

Reeves AL (1986) Beryllium. in: Friberg L, Nordberg GF, Vouk VB (Eds) Handbook on the toxicology of metals, 2nd edition, Elsevier, Amsterdam–New York–Oxford, 95–116

Rhoads K, Sanders CL (1985) Lung clearance, translocation, and acute toxicity of arsenic, beryllium, cadmium, cobalt, lead, selenium, vanadium, and ytterbium oxides following deposition in rat lung. Environ Res 36: 359–378

Sanders CL, Cannon WC, Powers GJ, Adee RR, Meier DM (1975) Toxicology of high-fired beryllium oxide inhaled by rodents. II. Metabolism and early effects. Arch Environ Health 30: 546–551

Sanderson WT, Ward EM, Steenland K, Petersen MR (2001) Lung cancer case-control study of beryllium workers, Am J Ind Med 39: 133–144

Schramel P, Wendler I, Angerer J (1997) The determination of metals (antimony, bismuth, lead, cadmium, mercury, palladium, platinum, tellurium, thallium, tin and tungsten) in urine samples by inductively coupled plasma-mass spectrometry. Int Arch Occup Environ Health 69: 219–223

Stiefel T, Schulze K, Zorn H, Tölg G (1980) Toxicokinetic and toxicodynamic studies of beryllium. Arch Toxicol 45: 81–92

Wegner R, Heinrich-Ramm R, Nowak D, Olma K, Poschadel B, Szadkowski D (2000) Lung function, biological monitoring, and biological effect monitoring of gemstone cutters exposed to beryls. Occup Environ Med 57: 133–139

WHO (World Health Organization) (1990) Beryllium. Environmental Health Criteria 106, Geneva

Zorn H, Stiefel TW, Beuers J, Schlegelmilch R (1988) Beryllium. in: Seiler HG (Ed.) Handbook on toxicity of inorganic compounds. Marcel Dekker Inc., New York–Basel, 105–114

Author: K.H. Schaller
Approved by the Working Group: 17.06.2002

1,4-Dichlorobenzene, Addendum

BAT/EKA	not yet established
Date of evaluation	2003

Since 1,4-dichlorobenzene was evaluated in Volume 3 of the present series (Greim and Lehnert 1998), the substance has been classified in Carcinogen category 2 and the MAK value has been withdrawn (Greim 2003). The BAT value has therefore also been withdrawn. 1,4-Dichlorobenzene has been listed in Section XIII.2 of the *List of MAK and BAT Values*. An EKA correlation is in evaluation.

References

Greim H (Ed.) (2003) 1,4-Dichlorobenzene. in: Occupational toxicants, Vol. 20, Wiley-VCH, Weinheim

Greim H, Lehnert G (Eds) (1998) 1,4-Dichlorobenzene. in: Biological exposure values for occupational toxicants and carcinogens – Critical data evaluation for BAT and EKA values, Vol. 3, Wiley-VCH, Weinheim, pp. 95–104

Mercury, organic mercury compounds

BAT/EKA	Withdrawn
Date of evaluation	2003

The organic mercury compounds are characterized chemically by covalent binding of the mercury to a carbon atom. At present of importance from an occupational-medical point of view are the salts of the phenylmercury compounds, methoxyethylmercury, the monoalkylmercury compounds, and methylmercury and ethylmercury salts. The different metabolic and toxicological behaviour of these mercury compounds makes it necessary to divide the substances into two classes. The organic compounds which metabolize in the organism to form inorganic mercury are of a toxicity similar to that of the inorganic mercury salts. These are phenylmercury and methoxyethylmercury compounds. The BAT values evaluated for inorganic mercury apply for these compounds. The much more stable alkylmercury compounds accumulate in the central nervous system and have greatly differing toxic effects compared to the above-mentioned rapidly metabolized organic mercury compounds.

The following occupational-medical, toxicological documentation applies for occupational exposure to alkylmercury compounds.

1 Metabolism and Kinetics

There are numerous studies available of the absorption, metabolism and excretion of alkylmercury in the human organism. Attention is drawn in particular to: WHO Criteria Documents "Mercury" (1976a, 1989) and Friberg, Nordberg and Vouk (1979).

At the workplace organic mercury compounds can be absorbed via inhalation, percutaneously or orally. The alkylmercury compounds are particularly volatile (a high saturation concentration, increasing from propylmercury to methylmercury compounds). Inhalation therefore plays the greatest role at the workplace. To date, there are no quantitative data available for the pulmonary absorption of mercury alkyls. It is assumed to be about 60 % of the inhaled amount (Task-group on metal accumulation 1973).

Organic mercury compounds, unlike the inorganic compounds, are absorbed in the intestine almost completely. For methylmercury, values of 80 % or more are given (Clarkson 1972). For radioactively labelled methylmercury other authors state absorption to be 95 % in man (Aberg *et al*. 1969, Miettinen 1973). There are no quantitative data for the cutaneous absorption of short-chain mercury alkyls in man. Intoxication after skin contact with methylmercury is, however, known (Suzuki *et al*. 1970).

The organic mercury compounds are distributed extremely rapidly and evenly in the organism. After some time, particularly in the case of methylmercury, there is a slow redistribution of the substance, predominantly in the brain (Berglund *et al.* 1971). 98 % of the total mercury in the brain is methylmercury.

The organic mercury compounds are readily soluble in lipids. They can affect the central nervous system severely and have a long biological half-time. They therefore tend to accumulate. The biological half-times in whole body and blood are around 70 days. A comprehensive description of accumulation and biological half-times can be found in the WHO Criteria Documents (1976a, 1989).

The faeces are the most important route of excretion for mercury after short-term or long-term absorption of alkylmercury. Studies with volunteers demonstrated that about 90 % of the absorbed mercury is excreted with the faeces (Aberg *et al.* 1969, Miettinen 1973).

2 Critical Toxicity

The pathological findings in persons who were exposed to organic mercury compounds have been comprehensively described by the Swedish Expert Group (1971) and Friberg and Vostal (1972). The findings show that alkylmercury compounds have mainly neurotoxic effects. These compounds are relatively stable and easily pass the blood-brain barrier. The most important pathological effects are the damage to nerve cells of the cortex, in particular in the region of the occipital cortex, and damage of differing extent in the granular layer of the cerebellum.

On the basis of the results of more recent animal experiments it can be assumed that ganglion cells of the dorsal spinal roots are more sensitive to methylmercury than nerve fibres and neurons of the dorsal spinal roots (Yip and Chang 1981).

Le Quesne *et al.* (1974) and von Burg and Rustam (1974) carried out electrophysiological investigations in patients after intoxication with methylmercury via contaminated food. The mercury levels in blood were in some of the patients between 13.8 and 87.8 µg/dl (no data given for the normal range). Despite clinically manifest neurological disturbances, both research groups found nerve conduction velocities and nerve action potentials in the normal range. On the basis of the results, they concluded that the nerve structures damaged by mercury seem to be located in the central nervous system, and here in particular in the region of the lower brain stem.

Usual nerve conduction velocities were also found by Rustam *et al.* (1975) and Snyder and Seelinger (1976) in both cases in 4 patients with severe intoxication with methylmercury.

There are neurophysiological and histological studies of the late changes in the sural nerve as a result of intoxication with organic mercury compounds ("Minamata disease") (Research Committee on Minamata Disease 1975, Miyakawa *et al.* 1976). The authors found a reduction in myelinated nerve fibres, with abnormal forms and regressive changes in Schwann's cells. They regard the lack of "onion skin formations" as a characteristic sign of intoxication with organic mercury. As a result of these histological anomalies, it could in their opinion thus be differentiated from other neuropathies.

Amin-Zaki et al. (1978) reported about intoxication with methylmercury in 49 Iraqi children. The mercury levels in blood were between 160 and 4800 µg/dl (no data given for the normal range). In some of the patients paraesthesia, paresis of different extent up to paralysis, and a pathological reflex status were clinically manifest.

Cinca et al. (1980) reported about 4 cases of intoxication with ethylmercury chloride via food. Electrophysiological investigations in 2 patients revealed nerve conduction velocity values in the normal range for motor fibres, while the sensory nerve conduction velocity of the median nerve and ulnar nerve was decreased to 26 and 42 m/s, respectively. After treatment, normal nerve conduction velocities were found at a check-up after 6 months.

Impairment of the kidneys after exposure to methylmercury compounds is very rare (Bakir et al. 1973). Therefore, for the evaluation of the BAT value, mainly the effects on the central nervous system must be discussed.

At present there is no known evidence of carcinogenic effects of organic mercury. In some studies chromosomal breaks and aberrations were reported. No statistical relationship was determined between the frequency of chromosomal breaks and the mercury levels in blood (WHO 1976a, Skerfving et al. 1974).

It is known from the intoxication epidemics in Japan and Iraq that short-chain mercury alkyls rapidly pass the placenta barrier. Both in Japan and Iraq, cases of intoxication were observed which were the result of prenatal exposure to methylmercury (WHO 1976a).

Reuhl and Chang (1979) described the effects of methylmercury on the development of the nervous system. On the basis of the results of animal experiments and cases of intoxication in man, it is clear that the foetus is more severely affected by methylmercury than the mother.

3 Exposure and Effects

Most of what we know about exposure and effect relationships results from studies about the alkylmercury intoxication epidemics in Iraq (Bakir et al. 1973, WHO 1976b) and Japan (Swedish Expert Group 1971) and studies of population groups who ate fish contaminated with mercury (Skerfving et al. 1974, Clarkson et al. 1973). A comprehensive description of the studies carried out there can be found in the WHO Criteria Document (WHO 1976a). These retrospective studies of cases of intoxication are the basis for the numerous dose–response relationships drawn up, the evaluation of threshold limit values for the highest permissible daily uptake of mercury and the critical organ concentrations in the brain and in the indicator media blood and hair.

If one looks at the dose–response relationships after alkylmercury intoxication, three main effects can be discerned: the neurotoxic, the teratogenic and the mutagenic effects. The two last effects are evidently only of importance for environmental intoxication epidemics. In the occupational-medical range the neurotoxic effects are most important.

The main signs and symptoms after intoxication with methylmercury or ethylmercury compounds are paraesthesia, hypaesthesia, ataxia, a narrowed visual field and impaired hearing. According to the Japanese findings, the effects of intoxication with alkylmercury are usually irreversible. On the other hand, there are reports that in Iraq improvement of the motor disturbances often occurred spontaneously; in addition, the paraesthesia was often reversible, while in Japan, however, it was permanent. In the WHO document (1976a) the results from the above-mentioned studies with population groups are listed in a table (Table 1). The given concentrations in hair and blood are the values for the most sensitive individuals in the affected population group. The results reveal that such "sensitive" persons can show symptoms of methylmercury intoxication at mercury concentrations of 20–40 µg/dl blood. The values in Table 1 are for neurological signs and symptoms in adults only.

Table 1. Mercury concentrations in blood and hair samples, and the amount of mercury in the body which caused effects (usually paraesthesia) in the most sensitive group of the population

Population	Total number of persons	Blood (µg/dl)	Mercury in the body (mg/kg)	References
Niigata/Japan	17	20–40	–	Swedish Expert Group 1971
Iraq	184	–	0.8	Shahristani *et al.* 1976
Iraq	125	24–48	0.5–0.8	Bakir *et al.* 1973
Iraq	427	–	0.7	Mufti *et al.* 1976

To date, no usable studies are known which place the levels of mercury in air in relation to the symptoms which can develop after exposure to alkylmercury compounds. In Sweden, a few cases of industrial exposure to ethylmercury and methylmercury were reported (Höök *et al.* 1954); in Japan one case of occupationally-related intoxication with ethylmercury was described (Kaisunumä *et al.* 1963) (see also Section 2). From these clinical observations the International Committee on Maximum Allowable Concentration of Mercury Compounds (ICMAC 1969) concluded that mercury levels in blood of more than 100 µg/dl may be accompanied by symptoms of intoxication. After industrial exposure to ethylmercury compounds, the suspected intoxication was accompanied by a mercury concentration in blood of 65 µg/dl (Kaisunumä *et al.* 1963). Popescu (1978) investigated three patients with occupationally-related alkylmercury intoxication. The levels of mercury excreted with the urine were found to be between 340 and 480 µg/l (no data given for the normal range). As a result of tremor and reduced sensitivity to temperature and pain, sensory peripheral neuropathy was suspected. In a follow-up examination 2 years later, the same neurological symptoms and psychological findings were detected, but to a reduced degree.

More recent investigations of persons exposed occupationally exclusively to organic mercury compounds are not known.

4 Selection of the Indicators

At present there are only exposure parameters available for monitoring persons occupationally exposed to organic mercury compounds.

Experimental studies in man and primates have shown that the mercury concentration in blood under steady-state conditions correlates linearly with the alkylmercury absorbed. In addition, there is a relationship between the mercury concentration in blood and the alkylmercury level in the critical organ, the brain. These relationships apply for non-toxic body burdens, too (Friberg et al. 1979). As more than 90% of the alkylmercury in blood is detectable in the erythrocytes, whole blood is the best indicator medium.

Also the level of mercury in hair can be used as an indicator. There are significant correlations with the levels of mercury in blood, the critical organ, and the body burden. Determination of this parameter has gained a certain importance in the environmental exposure range. For occupational-medical investigations, the determination of this parameter is not recommended for reasons of practicability. Responsible for this are mainly problems of contamination and the standardization of sampling, and the difficulty in interpreting the results. The amount of mercury excreted with the urine after exposure to alkylmercury is very small. The mercury concentrations in urine can be concealed by the inorganic mercury usually present. Determination of the mercury level in urine is, therefore, not a good indicator.

5 Methods

A comprehensive description of the analysis of mercury in biological material can be found in the BAT documentation for inorganic mercury in Volume 3 of the present series (Greim and Lehnert 1999). Other reviews can be found in the report about the environment and health criteria for mercury of the WHO (WHO 1976a, 1989) and in Schaller (1982).

The most suitable method is the determination of the mercury concentration in whole blood with the cold vapour technique. The quantitative determination of the level of mercury in blood can be carried out under the conditions of statistical quality control according to the guidelines of the German Federal Medical Council (Bundesärztekammer). Control material with different mercury concentrations is commercially available (Angerer et al. 1981).

6 Re-evaluation of the Biological Exposure Equivalents

In 1982 the BAT value for organic mercury compounds was set at 100 µg/l blood. The occupational-medical, toxicological documentation applied to occupational exposure to alkylmercury compounds. The database came mainly from environmental studies of accidental exposure in Japan and Iraq. In addition there were studies available of persons with high fish consumption. The data was evaluated by the WHO and the Swedish Expert Group (Swedish Expert Group 1971, WHO 1976).

At the time of the evaluation of the BAT value, the MAK value for alkylmercury compounds was 0.01 mg/m^3 measured as the inhalable fraction of the aerosol. In 1998 alkylmercury compounds and organic mercury compounds were classified in Carcinogen category 3 (corresponding to the current category 3B). The previous MAK value was withdrawn (Greim 2001). This re-classification of organic mercury compounds meant the BAT value had to be withdrawn.

6.1 Exposure and effects

There is more recent information available merely from studies of persons with high fish consumption and after the uptake of mercury from contaminated fish. The studies focus especially on children, as it is assumed that in these persons the CNS reacts particularly sensitively to exposure to organic mercury (Grandjean *et al.* 1999, Murata *et al.* 1999a, 1999b, Sörensen *et al.* 1999). These studies are not suitable for the evaluation of a workplace-specific value.

Dolbec *et al.* (2000) reports about investigations in a population group in the Brazilian Amazon region. The internal exposure to alkylmercury compounds was documented by the determination of mercury in hair and whole blood, where the total mercury levels were determined. The effects were taken into consideration e.g. in investigations of motor performance. Significant correlations were found between the total mercury levels in blood and hair ($r = 0.86$). In the hair investigations the first centimetre of hair was used. The ratio for the mercury concentration in blood to that in hair was calculated to be about 1:300. A concentration of mercury in blood of about 35 µg/l corresponded to a total mercury concentration in hair of 11 µg/g. According to data from the WHO 1990, a total mercury level in hair of 50 µg/g is associated with a risk of 5 % of neurological damage in adults. This observation is consistent with the investigations of a population in the Amazon region who consumed mainly fish (Lebel *et al.* 1998). Although hair analysis has proved useful in environmental studies of exposure to organic mercury, this parameter is not used in occupational-medical practice (Drexler and Schaller 2002). For the monitoring of persons occupationally exposed to alkylmercury compounds the determination of the total mercury level in whole blood is the parameter of choice.

6.2 Exposure equivalents for carcinogenic substances (EKA)

Recent studies of persons occupationally exposed to organic mercury compounds at the workplace which could be used for the evaluation of an EKA value are, to our knowledge, not available. This is, at least for the European region, also not to be expected, as the production of alkylmercury compounds for seed disinfectants or for other industrial uses has not taken place for many years.

As a consequence of the classification of organic mercury compounds in Carcinogen category 3B, the BAT value from 1982 has been withdrawn. Derivation of an EKA correlation (or a BLW value) on the basis of the data available is not possible.

7 References

Aberg B, *et al.* (1969) Arch Environ Health 19, 478–484
Amin-Zaki L, Majeed MA, Clarkson TW, Greenwood MR (1978) Brit Med J 1, 613–616
Angerer J, Schaller KH, Heinrich R (1981) Arbeitsmed Sozialmed Präventivmed 16, 125–127
Bakir F, *et al.* (1973) Science 181, 230–241
Berglund F, *et al.* (1971) Nord Hyg Tidskr Suppl 4, 106
Cinca I, Dumitrescu I, Onaca P, Serbanescu A, Nestorescu B (1980) J Neurol Neurosurg Psychiat 43, 143–149
Clarkson TW (1972) in: Goldberg l (Ed.) Critical reviews in toxicology 1, Cleveland, Ohio: CRC The Chemical Rubber Company 2, p. 203
Clarkson TW, *et al.* (1973) Arch Environ Health 26, 173–176
Drexler H, Schaller KH (2002) Haaranalysen in der klinischen Umweltmedizin. Dtsch Ärztebl 99: A 3026–3029 (Heft 45)
Dolbec J, Mergler D, Sousa Passos CJ, Sousa de Morais S, Lebel J (2000) Methylmercury exposure affects motor performance of a riverine population of the Tapajós river, Brazilian Amazon. Int Arch Occup Environ Health 73: 195–203
Friberg L, Nordberg GF, Vouk VB (1979) Handbook on the toxicology of metals. Elsevier/North Holland Biomedical Press
Friberg L, Vostal J (1972) Mercury in the environment. CRC Press, Cleveland, Ohio
Grandjean P, White RF, Nielsen A, Cleary D, de Oliveira Santos EC (1999) Methylmercury neurotoxicity in Amazonian children downstream from gold mining. Environ Health Perspect 107: 587–591
Greim H (Ed.) (2001) Mercury, organic compounds. in: Occupational toxicants, Vol. 15, Wiley-VCH, Weinheim
Greim H, Lehnert G (Eds) (1999) Mercury, metallic mercury and inorganic mercury compounds. in: Biological exposure values for occupational toxicants and carcinogens – Critical data evaluation for BAT and EKA values, Vol. 3, Wiley-VCH, Weinheim, pp. 123–142
Höök O, Lundgren KD, Swensson A (1954) Acta Med Scand 150, 131–137
ICMAC (1969) Arch Environ Health 19, 891–905
Kaisunumä H, *et al.* (1963) Report of the Japan Institute for Science of Labour No. 61
Lebel J, Mergler D, Branches F, Lucotte M, Amorim M, Larribe F, Dolbec J (1998) Neurotoxic effects of low-level methylmercury contamination in the Amazonian basin. Environ Res 79: 20–32
Le Quesne PM, Damluji SF, Rustam H (1974) J Neurol Neurosurg Psych 37, 333–339
Miettinen JK (1973) in: Miller MW, Clarkson TW (Eds) Mercury, mercurials and mercaptane, Springfield: CC Thomas, p. 233

Miyakawa T, Murayama E, Sumiyoshi S, Deshimaru M, Fujimoto T, Hattori E, Shikai I (1976) Acta Neuropath (Berlin) 36, 131–138

Mufti AW, *et al.* (1976) in: WHO conference on intoxication due to alkylmercury treated seed, Baghdad 9–13 November 1974, Geneva, World Health Organization, p. 23 (Suppl to Bull WHO, Vol. 53)

Murata K, Weihe P, Araki S, Budtz-Jörgensen E, Grandjean P (1999a) Evoked potentials in Faroese children prenatally exposed to methylmercury. Neurotoxicol Teratol 21: 471–472

Murata K, Weihe P, Renzoni A, Debes F, Vasconcelos R, Zino F, Araki S, Jörgensen PJ, White RF, Grandjean P (1999b) Delayed evoked potentials in children exposed to methylmercury from seafood. Neurotoxicol Teratol 21: 343–348

Popescu HI (1978) Brit Med J 1, 613–616

Research Committee on Minamata Disease (1975) Pathological, clinical and epidemiological research about Minamata Disease, 10 years after, Kumamoto, Kumamoto University

Reuhl KR, Chang LW (1979) Neurotoxicology 1, 21–55

Rustam H, von Burg R, Amin-Zaki L, Hassani SE (1975) Arch Environ Health 30, 190–195

Schaller KH (1982) Staub-Reinhaltung der Luft 42, 142

Shahristani H, *et al.* (1976) in: WHO conference on intoxication due to alkylmercury treated seed, Baghdad 9–13 November 1974, Geneva, World Health Organization, p. 105 (Suppl to Bull WHO, Vol. 53)

Skerfving S, *et al.* (1974) Environ Res 7, 83–98

Snyder RD, Seelinger DF (1976) J Neurol Neurosurg Psychiat 39, 701–704

Sörensen N, Katsuyuki M, Budtz-Jörgensen E, Weihe P, Grandjean P (1999) Prenatal methylmercury exposure as a cardiovascular risk factor at seven years of age. Epidemiology 10: 370–375

Suzuki T, Miyama T, Katsunuma H (1970) Industr Health 8, 39–47

Swedish Expert Group (1971) Nordisk Hygienisk Tidsskrift Suppl 4, 65

Task-group on metal accumulation (1973) Environ Physiol Biochem 3, 65–107

Von Burg R, Rustam H (1974) Electroenc Clin Neurophysiol 37, 381–392

WHO (World Health Organization) (1976a) Mercury. Environmental Health Criteria 1, Geneva

WHO (1976b) Conference on intoxication due to alkylmercury treated seed, Baghdad, 9–13 November 1974, Geneva, World Health Organisation (Suppl to Bull WHO, Vol. 53)

WHO (World Health Organization) (1989) Mercury – Environmental aspects. Environmental Health Criteria 86, Geneva

WHO (World Health Organization) (1990) Methylmercury. Environmental Health Criteria 101, Geneva

Yip RK, Chang LW (1981) Environ Res 26, 152–167

Author: K.H. Schaller
Approved by the Working Group: 23.07.2003

BLW Value Documentation

Arsenic and inorganic arsenic compounds

(with the exception of arsenic hydride and its salts)

BLW		50 µg arsenic and methylated metabolites/ l urine[1]	
		Sampling time: end of exposure or end of shift after several previous shifts	
Date of evaluation		2002	

Substance	CAS No.	Formula	Molecular weight	Melting point (°C)
arsenic	7440-38-2	As	74.92	817
arsenic trioxide	1327-53-3	As_2O_3	197.82	193
arsenous acids	36465-76-6	$HAsO_2$ and	125.94	not stated
	13464-58-9	H_3AsO_3		
and their salts, e.g.[2]				
sodium arsenite	7784-46-5	$Na\,AsO_2$	191.89	not stated
arsenic pentoxide	1303-28-2	As_2O_5	229.82	315
arsenic acid	7778-39-4	H_3AsO_4	141.94	35
and its salts, e.g.				
lead arsenate	7778-39-4	$Pb_3(AsO_4)_2$	899.43	382
calcium arsenate	7778-44-1	$Ca_3(AsO_4)_2$	398.07	1

MAK [last established: 1971, 2002]	Carcinogen category 1

(See also the documentation for carcinogenic substances for arsenic trioxide in Vol. 2)

Arsenic, arsenic trioxide and pentoxide, arsenous acids, arsenic acid and their salts (arsenites and arsenates) are carcinogenic substances (Carcinogen category 1) and do not, therefore, have MAK and BAT values (Greim 2002). The former TRK value (technical exposure limit) for arsenic and its inorganic compounds was 0.10 mg arsenic/m^3 (measured as the inhalable fraction of the aerosol; calculated as arsenic). This arsenic concentration in air correlated with an arsenic concentration in urine of 130 µg/l at the end of exposure or the end of the shift after several previous shifts (see the EKA documentation for arsenic trioxide in Volume 2 of this series; Greim and Lehnert 1995). The inorganic arsenic and methylated metabolites are analysed together.

---

[1] volatile arsenic compounds determined by means of direct hydrogenation with hydride-AAS
[2] with the exception of gallium arsenide

The MAK-Collection Part II: BAT Value Documentations, Vol. 4. DFG,
Deutsche Forschungsgemeinschaft
Copyright © 2005 WILEY-VCH Verlag GmbH & Co. KGaA, Weinheim
ISBN: 3-527-27049-3

In addition to their carcinogenic effects in man, arsenic and the above-named arsenic compounds have toxic effects on the skin, the nervous system and the vascular system. Even with internal exposure values to arsenic below those corresponding to external exposure to arsenic trioxide at the level of the TRK value (EKA_{TRK}), such toxic effects have been described, so that with exposure at the level of the EKA_{TRK} value protection against these effects is not guaranteed (Wichmann and Lehnert 1987).

1 Metabolism and Kinetics

1.1 Absorption and distribution

Man is exposed to different forms of organic and inorganic arsenic in foodstuffs, water and other matrices. Studies of the kinetics and metabolism of arsenic and its compounds are, therefore, complex. Inhalation exposure to inorganic arsenic occurs mainly at the workplace. In air, arsenic and its inorganic compounds are particle-bound and respiratory absorption therefore takes place in a two-phase process. After deposition of the particles in the respiratory tract and lungs, the arsenic is desorbed from the deposited particles. The extent to which the particles are deposited and the resulting exposure to arsenic therefore also depend on the size of the particles. Desorption of the arsenic compounds from the particles, however, depends on their solubility in water. For the reasons named, pulmonary retention cannot be estimated or quantified with sufficient certainty.

All water-soluble inorganic arsenic compounds are absorbed after ingestion to a considerable extent (up to 95 %) in the gastrointestinal tract (Marquardt 1997). The bioavailability depends, however, among other things also on the matrix of the foodstuff. Dermal absorption can be regarded as low compared to absorption after inhalation and gastrointestinal absorption (Wester *et al.* 1993).

Inhaled and ingested inorganic arsenic compounds first of all enter the blood. With a half-time of two hours, inorganic arsenic is rapidly eliminated from the blood. The arsenic compounds are distributed, as investigations with radioactively labelled arsenic compounds showed, in all the organs investigated. In addition to renal elimination, biliary excretion takes place. Experiments showed that elimination occurs in three phases in man. About 66 % of the administered doses is renally eliminated with a half-time of 2.1 days, around 30 % with a half-time of 9.4 days and the rest (4 %) with a half-time of 38.4 days (Marquardt 1997).

1.2 Metabolism

The metabolism of arsenic in man and in many species of animal takes place in two steps; first of all arsenic(V) is usually reduced to arsenic(III), which is then methylated to form monomethylated and dimethylated arsenic compounds. A detailed overview of the metabolism of arsenic compounds can be found in the IPCS programme (WHO 2001) and in the EKA documentation for arsenic trioxide in Volume 2 of this series (Greim and Lehnert 1995).

2 Critical Toxicity

A detailed description of the symptoms of acute and chronic intoxication with arsenic can be found in the current WHO monograph on arsenic and arsenic compounds (WHO 2001).

2.1 Acute toxicity

The symptoms of intoxication after short-term and long-term absorption of inorganic arsenic compounds are similar and depend on the dose and bioavailability of the substance. The trivalent arsenic compounds are three to four times more toxic than the corresponding pentavalent compounds. The oral LD_{50} values for inorganic arsenic compounds, depending on the arsenic species and the experimental animal, are in the range from 7 to 100 mg/kg body weight (Marquardt 1997).

The clinical symptoms after the ingestion of arsenic compounds occur within 30 to 60 minutes. Described are cardiovascular collapse, CNS depression and severe gastrointestinal symptoms. Death takes place within a few hours. With less dramatic cases of intoxication the main symptoms are gastrointestinal complaints including a metallic taste, a dry mouth, burning lips, dysphagia, vomiting attacks and occasional haematemesis.

2.2 Chronic toxicity

The carcinogenic effects of inorganic arsenic compounds have been demonstrated in epidemiological studies in man. In addition to tumours of the bronchi and the lungs, in particular the skin tumours (basaliomas situated on the trunk) are pathognomonic.

In addition to the carcinogenic effects, inorganic arsenic compounds cause damage in particular to the cardiovascular system and the nervous system. Long-term exposure to inorganic arsenic compounds is associated with cardiovascular diseases and cerebrovascular impairment. In particular the impairment of the peripheral vascular muscle-coat, even as far as gangrene of the affected extremities, has been consistently observed (WHO 2001). The neurotoxic effects can affect both the peripheral nervous system (Feldman et al. 1979, WHO 2001) and the central nervous system (WHO 2001).

Arsenic and its compounds are sensitizing to the skin only in exceptional cases and the described association between exposure to arsenic and diabetes mellitus needs further clarification (Rahman et al. 1995).

3 Exposure and Effects

Peripheral vascular diseases were described in persons who consumed spring water with a high arsenic concentration. Symptoms such as acrocyanosis and Raynaud's phenomenon were described after cumulative absorption of 8 g arsenic (Borgono et al. 1977). After cumulative absorption of 20 g arsenic, gangrene of the feet developed (Pershagen and Vahter 1979).

In occupationally exposed copper smelters, vasospastic changes in the arteries of the fingers were reported (Lagerkvist et al. 1986, 1988). The workers were mainly exposed to arsenic trioxide. The exposure was determined by analysing the inorganic arsenic and its metabolites in urine. The mean arsenic concentration in the urine of the exposed workers was 71 µg/l (range 10–340 µg/l), that of the asymptomatic controls 7 µg/l (range 5–20 µg/l). There were statistically significant differences between the exposed persons and the controls in the systolic blood pressure of the finger after cooling and the anamnesis of Raynaud's syndrome. It cannot be excluded, however, that in the past the exposure of the workers was much higher than at the time of the investigation. In the same group exposed to arsenic, the peripheral nervous system was investigated by means of electromyography and determination of the sensory nerve conduction velocity in the arms and legs. The exposed workers were found to have a slightly reduced nerve conduction velocity, which was interpreted as a subclinical sign of neuropathy (Blom et al. 1985). In another study, an increased incidence of clinically manifest and subclinical neuropathy in copper smelters was reported when the arsenic concentration in urine reached 250 µg/l (Feldman et al. 1979). One publication describes peripheral neuropathy in 51 persons with arsenic concentrations in urine above 100 µg/l (Heyman et al. 1956). Engel et al. (1994) give an overview of the vascular effects after long-term exposure to arsenic in their review based on 177 publications.

In studies of the nephrotoxic (Foà et al. 1987) and hepatotoxic effects (Kodama et al. 1976), statistically significant changes were found in exposed workers who excreted mean concentrations of arsenic in the urine of 102 µg/l and 82 µg/l, respectively.

To conclude, there are sufficient data from occupational-medical studies (see Table 1) which show adverse effects can occur at relatively low arsenic concentrations in urine, i.e. even below the former EKA_{TRK} of 130 µg/l.

Table 1. Relationship between adverse effects on health and the arsenic concentrations in occupationally exposed persons (modified from ACGIH (2001)

Observed effects	Arsenic in urine (µg/l) median (range)	Author
vasospastic tendency	71 (10–340)	Lagerkvist et al. 1986
peripheral neuropathy	71 (10–340)	Blom et al. 1985
peripheral neuropathy	< 250	Feldman et al. 1979
peripheral neuropathy	> 100	Heyman et al. 1956
impaired kidney function	103	Foà et al. 1987
increase in liver enzymes	82 ± 49 (SD)	Kodama et al. 1976

4 Selection of the Indicators

After exposure to inorganic arsenic there is a significant and specific increase in the arsenic species monomethylarsenic acid, dimethylarsinic acid and inorganic arsenic excreted in urine. The determination of these three arsenic species in urine is therefore the preferred method for the biological monitoring of workers exposed to inorganic arsenic. The determination of these arsenic species is not influenced by organic arsenic compounds, such as arsenobetaine and arsenocholine, taken up with food. Determination of the total amount of arsenic excreted in urine is not suitable for monitoring investigations because organic arsenic compounds of maritime origin are also determined (see the EKA documentation for arsenic trioxide in Volume 2 of this series; Greim and Lehnert 1995).

Inorganic arsenic and its metabolites are rapidly excreted after the beginning of exposure. The arsenic concentration in urine increases slowly and remains at a relatively constant level during the first three days of exposure. During the working day and from the end of work to the beginning of the next shift there are no notable changes in the concentration. The elimination kinetics show that during the working week there is significant accumulation of arsenic and its metabolites. Sampling should therefore take place at the end of the working week (see Section 1.1).

The determination of arsenic in blood and the arsenic concentrations in hair and finger nail samples have not acquired any importance in the biomonitoring of occupationally exposed persons (see the EKA documentation for arsenic trioxide in Volume 2 of this series; Greim and Lehnert 1995).

5 Methods

Inorganic and methylated organic arsenic compounds in urine are transformed for quantitative analysis into volatile hydrides and determined by atomic absorption spectroscopy. As a result of this method of direct hydrogenation of the urine sample only inorganic arsenic compounds and their metabolites monomethylarsenic acid and dimethylarsinic acid are transformed into volatile hydrides and quantified. A tested method can be found in "Analyses of Hazardous Substances in Biological Materials" (Angerer and Schaller 1991). If the urine sample is digested wet or dry the total arsenic content is determined. Therefore preparation of the urine samples should on no account be carried out. The following refers to the determination of inorganic arsenic and its methylated metabolites. In biological monitoring this arsenic fraction should be determined (Angerer and Schaller 1991).

6 Background Exposure

In persons not occupationally exposed, the diet is the main route of absorption for inorganic arsenic compounds. Food, with the exception of sea-food, contains less than 0.25 mg arsenic/kg; various kinds of fish between 1 and 10 mg/kg, and shell fish more than 100 mg/kg. In food of maritime origin the arsenic is in organic form, usually as arsenobetaine and arsenocholine. The daily uptake with food is estimated to be between 0.04 (without fish) and 0.19 mg arsenic (with fish). In comparison, in Germany, insignificant amounts of arsenic are taken up from the drinking water and air. With the hydride-AAS method suggested in Section 5 the aromatic arsenic compounds excreted unmetabolized in the urine are not determined. Under these analytical conditions the background exposure is below 25 µg/l urine. Recent investigations have shown, however, that also the level of excretion of the named arsenic fraction (inorganic arsenic, monomethylarsonic acid and dimethylarsinic acid) can be influenced by the consumption of large amounts of food of maritime origin. These foods can already contain the methylated substances, in particular dimethylarsinic acid (Apostoli *et al.* 1999). It is therefore recommended that in future only the inorganic arsenic fraction be determined. Separation of the arsenic species is, however, a time-consuming and complicated analytical procedure. Reference values for the various arsenic species in urine samples from the general population of northern Germany have been published (Heinrich-Ramm *et al.* 2001).

7 Evaluation of the BLW Value

The former TRK value for arsenic trioxide and pentoxide, arsenous acids, arsenic acid and their salts was 0.10 mg arsenic/m^3 air. In 1994 an EKA correlation was established for arsenic trioxide, in which an arsenic concentration in the air of 0.10 mg/m^3 corresponds to the excretion of 130 µg arsenic/l urine. The TRK value was not toxicologically grounded and observation of the value did not provide sufficient protection against the toxic effects named in Section 2.2.

For the prevention of (non-carcinogenic) toxic effects of arsenic, irrespective of the carcinogenic potency of arsenic, exposure should remain below 50 µg arsenic/l urine. Taking into account the literature cited, it is to be expected that arsenic-induced effects can develop at arsenic concentrations in urine in the range of 100 µg/l. Changes in the peripheral nerves of occupationally exposed persons were detected even at a mean arsenic concentration of 71 µg/l urine (Blom *et al.* 1985). The BLW value has therefore been set at

50 µg arsenic/l urine

This value is also compatible with the American BEI value (Biological Exposure Indices) (35 µg arsenic/l urine, sampling at the end of the working week), as this refers to the collective median and does not represent an individual ceiling value (ACGIH 2001).

8 Interpretation of Data

The toxicologically relevant arsenic concentration in urine is determined with the method described in "Analyses of Hazardous Substances in Biological Materials" (Angerer and Schaller 1991). The arsenic taken up with food of maritime origin is not determined. The use of the so-called hydride method is therefore imperative. The total amount of arsenic renally excreted must not be determined after preparation of the urine sample as the results can be greatly influenced by the food consumed. Fish and other sea-food can contain a high level of organic arsenic compounds. These compounds are rapidly absorbed by the gastrointestinal tract and can lead to the increased excretion of arsenic in urine. Arsenic concentrations in urine in the range of the BLW value can be reached and even exceeded (Mindt-Prüfert *et al.* 1999).

It should also be noted that also with the use of the hydride-AAS method, the excretion of inorganic arsenic and its methylated metabolites can be influenced by high fish consumption (Apostoli *et al.* 1999). Fish can also frequently contain methylated arsenic compounds (in particular dimethylarsinic acid). The total amount of arsenic in urine can, therefore, be increased.

If the BLW is exceeded, the individual should be specifically asked about his fish consumption over the last few days, and if necessary the arsenic concentration should be redetermined after a period without fish. After a single meal containing sea-food contaminated with organic arsenic compounds, the amount of arsenic excreted within 24 hours was 50 % to 70 % of the total amount of arsenic taken up (Marquardt 1997). Short-term control investigations can therefore be useful.

In some countries, such as Bangladesh, Argentina, Chile and India, considerable amounts of arsenic are taken up by the general population with the drinking water from contaminated wells. This causes amounts of arsenic to be excreted in urine which are above the BLW value (WHO 2001).

9 References

ACGIH (American Conference of Governmental Industrial Hygienists) (2001) Arsenic and soluble inorganic compounds: BEI® 7th edition, Documentation. Publication # 7DOC-665, ACGIH®, Cincinatti

Angerer J, Schaller KH (1991) Arsenic. in: Henschler D (Ed.) Analyses of hazardous substances in biological materials, Vol. 3, VCH, Weinheim

Apostoli P, Bartoli D, Alessio L, Buchet JP (1999) Biological monitoring of occupational exposure to inorganic arsenic. Occup Environ Med 56: 825–832

Blom S, Lagerkvist B, Linderholm H (1985) Arsenic exposure to smelter workers. Clinical and neurophysiological studies. Scand J Work Environ Health 11: 265–269

Borgono JM, Vincent P, Venturino H, Infante A (1977) Arsenic in the drinking water of the city of Antofagasta: epidemiological and clinical study before and after the installation of a treatment plant. Environ Health Perspect 19: 103–105

Engel RR, Hopenhayn-Rich C, Receveur O, Smith AH (1994) Vascular effects of chronic arsenic exposure: a review. Epidemiol Rev 16: 184–209

Feldman RG, Niles CA, Kelly-Hayes M, Sax DS, Dixon WJ, Thompson DJ, Landau E (1979) Peripheral neuropathy in arsenic smelter workers. Neurology 29: 939–944

Foà V, Colombi A, Maroni M, Barbieri F, Franchini I, Mutti A, De Rosa E, Bartolucci GB (1987) Study of kidney function of workers with chronic low level exposure to inorganic arsenic. in: Foà V, Emmett EA, Maroni M, Colombi A (Eds) Occupational and environmental chemical hazards, Cellular and biochemical indices for monitoring toxicity, Ellis Horwood Limited Publishers, Chichester, 362–367

Greim H (Ed.) (2002) Arsen und anorganische Arsenverbindungen (mit Ausnahme von Arsenwasserstoff). Gesundheitsschädliche Arbeitsstoffe, Toxikologisch-arbeitsmedizinische Begründungen von MAK-Werten, 35. Lieferung, Wiley-VCH, Weinheim

Greim H, Lehnert G (Eds) (1995) Arsenic trioxide. in: Biological exposure values for occupational toxicants and carcinogens – Critical data evaluation for BAT and EKA values, Vol. 2, VCH, Weinheim, pp. 123–136

Heinrich-Ramm R, Mindt-Prüfert S, Szadkowski D (2001) Arsenic species excretion in a group of persons in northern Germany – Contribution to the evaluation of reference values. Int J Hyg Environ Health 203: 475–477

Heyman A, Pfeiffer JB Jr., Willett RW, Taylor HM (1956) Peripheral neuropathy caused by arsenical intoxication. A study of 41 cases with observations on the effect of BAL (2,3, dimercapto-propanol). N Engl J Med 254: 401–409

Kodama Y, Ishinishi N, Kunitake E, Inamasu T, Nobutomo K (1976) Subclinical signs of the exposure to arsenic in a copper refinery. in: Nordberg GF (Ed.) Effects and dose-response relationships of toxic metals, Elsevier Scientific Publishing Company, Amsterdam, 464–470

Lagerkvist B, Linderholm H, Nordberg GF (1986) Vasospastic tendency and Raynaud's phenomenon in smelter workers exposed to arsenic. Environ Res 39: 465–474

Lagerkvist BE, Linderholm H, Nordberg GF (1988) Arsenic and Raynaud's phenomenon. Vasospastic tendency and excretion of arsenic in smelter workers before and after the summer vacation. Int Arch Occup Environ Health 60: 361–364

Marquardt H (Ed.) (1997) Lehrbuch der Toxikologie, Spektrum Akademischer Verlag, Heidelberg–Berlin, 510

Mindt-Prüfert S, Heinrich-Ramm R, Szadkowski D (1999) Zum alimentären Einfluß auf die Arsenausscheidung im Urin bei arbeitsmedizinischen Vorsorgeuntersuchungen. in: Rettenmeier AW, Feldhaus C (Ed.) Dokumentationsband über die 39. Jahrestagung der Deutschen Gesellschaft für Arbeitsmedizin und Umweltmedizin e.V., Druckerei Rindt, Fulda, 591–593

Pershagen G, Vahter M (1979) Arsenic: A toxicological and epidemiological appraisal. 071-SNV Rapporter från SNV, Statens naturvårdverk, Solna, Sweden

Rahman M, Wingren G, Axelson O (1995) Diabetes mellitus among Swedish art glass workers – an effect of arsenic exposure? Scand J Work Environ Health 22: 146–149

Wester RC, Maibach HI, Sedik L, Melendres J, Wade M (1993) *In vivo* and *in vitro* percutaneous absorption and skin decontamination of arsenic from water and soil. Fundam Appl Toxicol 20: 336–340

WHO (World Health Organization) (2001) Arsenic and arsenic compounds. Environmental Health Criteria 224, Geneva

Wichmann N, Lehnert G (1987) Arbeitsmedizinische Erfahrungen und TRK-Wert für Arsen. Arbeitsmed Sozialmed Präventivmed 22: 18–20

Author: H. Drexler
Approved by the Working Group: 10.02.2002

Cresols (all isomers)

BLW	200 mg total cresol/l urine
	Sampling time: end of exposure or end of shift
Date of evaluation	2003

Chemical name	Synonyms	CAS No.
Isomer mixture		
Methylphenol	Hydroxytoluene	1319-77-3
Individual isomers		
2-Methylphenol	*o*-Cresol, *o*-cresylic acid, 1-hydroxy-2-methylbenzene, 2-hydroxytoluene	95-48-7
3-Methylphenol	*m*-Cresol, *m*-cresylic acid, 1-hydroxy-3-methylbenzene, 3-hydroxytoluene	108-39-4
4-Methylphenol	*p*-Cresol, *p*-cresylic acid, 1-hydroxy-4-methylbenzenel, 4-hydroxytoluene	106-44-5

Formula	C_7H_8O
Molecular weight	108.14
Melting point	11.5–34.8 °C
Boiling point	191–202.2 °C
Vapour pressure at 25 °C	0.15–0.39 hPa
Density at 20 °C	1.02–1.03 g/cm^3
Odour threshold	0.0012–22 mg/m^3
MAK [last established: 2000]	Carcinogen category 3A

Cresols are colourless to yellowy, oily, poorly water-soluble liquids or crystals with a pungent, tar-like smell, which are of low volatility, are hardly inflammable and burn to produce large amounts of soot. The vapour is much heavier than air. Cresol–air mixtures may explode at high temperatures.

180   *Cresols (all isomers)*

The isomeric cresols are starting materials for the production of plastics, tanning agents, dyes, disinfectants and perfumes, insecticides and herbicides. In addition, they are used for the production of cresol resins, of solvents for wire coatings, for the production of phosphoric acid esters, hydraulic fluids and of cresyl compounds (ATSDR 1992, Calabrese 1986, Deichmann and Keplinger 1981, Greim 2000a, WHO 1995).

1 Metabolism and Kinetics

1.1 Absorption and distribution

The isomeric cresols are absorbed mainly via the skin and respiratory tract (ATSDR 1992). In analogy to phenol, high levels of absorption are to be expected. As a result of the low volatility and intensive odour (warning signal) of the substance, the danger of acute intoxication from inhalation is, however, small under normal circumstances.

As a result of rapid local anaesthesia and the lack of pain as a warning signal, skin contact often remains unnoticed. Absorption of acutely toxic doses of the substance can therefore take place, and even of lethal doses if large areas of skin are wetted; this is regarded as the main source of danger when handling the substance occupationally (ATSDR 1992).

After ingestion, the substance is quickly absorbed in the gastrointestinal tract, leading rapidly to systemic effects (ACGIH 2001, ATSDR 1992, Deichmann and Keplinger 1981).

1.2 Metabolism and elimination

Percutaneously absorbed and inhaled cresol is rapidly detoxified, mainly by conjugation (sulfate, glucuronide), and excreted by the kidneys. In general, no free cresol can be detected in plasma, even after acute intoxication. Depending on their hydrolysis by intestinal bacteria, conjugates eliminated in the bile enter enterohepatic circulation (ATSDR 1992, Deichmann and Keplinger 1981).

According to the results of animal experiments, for *p*-cresol there is another metabolic pathway that leads via the oxidation of the methyl group to the formation of *p*-hydroxybenzoic acid. A reactive quinone-methide intermediate is formed, which can be bound to glutathione or macromolecules. Such compounds, which are known from the metabolism of benzene (see the chapter on "Phenol" in Volume 1 of this series; Henschler and Lehnert 1994), have been detected to date only for *p*-cresol (DECOS 1998, Thompson *et al.* 1996).

For the other cresol isomers, the extent of ring hydroxylation resulting from the formation of dihydroxytoluenes seems to be small (Deichmann and Keplinger 1981).

Some of the *p*-cresol detected in the urine and faeces is formed during the bacterial breakdown of amino acids in the intestine. Furthermore, all cresol isomers are formed as metabolites of toluene (Greim 2000a, Nise 1992, Sherwood and Carter 1970).

2 Critical Toxicity

2.1 Acute toxicity

The slight differences in the toxicity of the cresol isomers (*p* > *o* > *m*-cresol) are, from a practical point of view, negligible. Their effects, also as a mixture of isomers, are similar to those of phenol; the local effects on the skin and mucous membranes are, however, slightly less pronounced (ATSDR 1992, Greim 2000a).

After direct contact of the eyes with the substance there is the danger of severe damage to the eyes, and even of blindness. Undiluted cresol caused clouding and vascularization in the rabbit eye (ATSDR 1992).

The cresols and their concentrated solutions are caustic to the skin, and if they penetrate deeply also cause damage to the vessels. First of all a burning sensation is felt, which rapidly develops into a local loss of feeling. Finally, white to dark-coloured tissue damage can occur, and even gangrene (ATSDR 1992, Moeschlin 1972).

Diluted solutions cause reddening of the skin and the formation of blisters. In hypersensitive persons inflammation of the skin was induced by solutions with cresol concentrations as low as < 0.1 % (Moeschlin 1972).

The vapour given off when cresol is heated has irritative effects on the respiratory tract, can damage the lungs and can lead, usually with a delay, to pulmonary oedema.

Ingestion of cresol causes painful local effects in the mucous membranes affected; whitish slough is formed, the person has difficulty in swallowing, sometimes oedema of the glottis develops, and nausea and stomach ache are reported. As a result of the local paralysis of the sensitive nerves of the stomach, vomiting is usually not observed even after acute intoxication with cresol (Moeschlin 1972).

The main acute systemic effects, especially after contact with the substance of large areas of skin or ingestion of more than 1 g cresol, are a rapid impairment of the central nervous and cardio-vascular systems. The uptake of more than about 10 g cresol is generally lethal (ATSDR 1992, WHO 1995).

The symptoms observed during the generally slow course of intoxication are: headaches, dizziness, usually a slower pulse and a drop in blood pressure, tinnitus, visual disorders, breathing difficulties, muscle weakness, and sometimes convulsions and impaired consciousness, often leading to unconsciousness. Damage to the blood cells and kidneys has been observed, including the formation of Heinz bodies, methaemoglobinaemia, haemolysis, haematuria, proteinuria, oliguria and sometimes also anuria. The haematological effects were observed mainly in persons with deficient glucose-6-phosphate dehydrogenase (G-6-PDH) activity (Chan *et al.* 1971, Moeschlin 1972, Roots *et al.* 1992).

After some time, further complications, such as bronchopneumonia, damage to the pancreas and liver, and strictures of the oesophagus and intestine can develop. A rapid course of intoxication is possible after the uptake of high doses. Usually a severe state of collapse is observed, which can lead to death as a result of respiratory arrest, sometimes after 10 to 60 minutes, but also immediately (ATSDR 1992, Moeschlin 1972).

As a result of a polymorphism of the isoenzymes of UDP-glucuronosyl transferase (UDPG), around 6% of Central Europeans can glucuronidize only slowly (see the documentation on "Phenol" in this series; Henschler and Lehnert 1994). In some rare cases this defect is amplified by the lack of phosphoadenosine-phosphosulfate (PAPS) sulfotransferase responsible for sulfation. In these persons, in particular accidental exposure to phenol or cresols generally leads to the increased formation of free phenol and cresols, which can cause massive impairments in kidney function and even acute kidney failure (ATSDR 1992, Calabrese 1986, Lewalter and Neumann 1998, Lewalter et al. 1993).

These adverse effects have been extensively investigated in the case of phenol (see the documentation on "Phenol" in this series; Henschler and Lehnert 1994), but there are only case reports available for the cresols.

After the accidental exposure to *m*-cresol of a UDPG-deficient worker, the data for exposure and effects shown in Table 1 were determined during monitoring of the course of intoxication.

Table 1. Case report of intoxication with *m*-cresol of a UDPG-deficient worker (Lewalter, personal communication to the Commission)

Time period after the accident	Urine		Serum
	Free *m*-cresol [µg/l urine]	Total *m*-cresol [mg/l urine]	Creatinine [mg/dl serum]
1 hour	120	226	1.0
12 hours	70	99	2.2
24 hours	30	34	5.0
72 hours	10	6	2.5
3 months (occupational-medical check-up)	< 1.0	0.001	1.0

The slow increase in serum creatinine after acute intoxication with cresol is unusual.

2.2 Chronic toxicity

As seen in man and experimental animals, long-term exposure to cresols affects the same organs which are damaged after short-term exposure, in particular the liver.

In persons who inhaled the substance, headaches, nausea, vomiting, increased blood pressure, renal functional disorders, marked tremor and irritation of the upper airways were observed (ATSDR 1992, Deichmann and Keplinger 1981, Moeschlin 1972). A biochemical change described is the disturbance of the blood–calcium homeostasis. Other possible symptoms are difficulties in swallowing, the increased formation of saliva, diarrhoea, dizziness and impaired consciousness. The prognosis in particular after severe damage to the liver and kidneys is bad; in some cases this can even lead to death.

Cresols can cause discoloration and, in toxic doses, inflammation of the skin.

The genotoxicity studies which have been carried out with the individual cresol isomers *in vivo* have not yielded any evidence of genotoxic effects. A suggestion that the substances could have genotoxic effects in mammalian cells may be deduced from the positive results of some *in vitro* studies, but the relevance of these studies is not always clear (Greim 2000a). It is known that *p*-cresol is converted metabolically to a reactive derivative with a quinoid structure; for *o*-cresol such activation is also theoretically possible. *m*-Cresol, which cannot be activated metabolically in this way, yielded negative results in all genotoxicity studies. The results are thus largely consistent and compatible with a weak genotoxic effect, especially of *p*-cresol (Greim 2000a).

In initiation-promotion studies the cresol isomers reveal a promoting effect like that of phenol. As there is evidence of promoting effects which are independent of the skin irritation, in 1999 the cresol isomers were classified in Carcinogen category 3 Section III of the *List of MAK and BAT Values* (Greim 2000a). In 2000 they were reclassified in the new category 3A (Greim 2000b).

3 Exposure and Effects

3.1 Relationship between external and internal exposure

In the case of the cresol isomers external exposure is any kind of exposure that can lead to inhalation or percutaneous absorption of the substance.

In occupational-medical practice, the excretion of cresols after contact with the substance is often evaluated on the basis of so-called "empirical values" (Lewalter and Neumann 1998). The relationships listed in Table 2 between the external and internal exposure to total cresols are thereby observed (Lewalter, personal communication to the Commission). In the meantime these correlations have been supplemented for the low dose range by a study with coking plant workers (Bieniek 1997).

3.2 Relationship between internal exposure and effects

The internal exposure is the biologically relevant concentration of the cresols above all in the liver and kidneys. Like with phenol, also after increasing exposure to cresols an increase is observed in the free cresols in urine, which in UDPG-deficient workers can be diagnosed even after markedly lower cresol levels (Table 1) (see also the documentation on "Phenol" in this series; Henschler and Lehnert 1994).

The exposure parameters are the concentrations of total cresols and free cresols in body fluids, such as blood and urine, under the specific pharmacokinetic conditions.

Table 2. Data for the relationship between external and internal exposure after occupational contact with cresols

Number of persons [N]	Air Total cresols (mean shift values) [ml/m^3]	[mg/m^3]	Urine Total cresols [mg/l urine]	References
75	0.05	0.22	19	Bieniek 1997
520	0.91	4	68 (36–103)	Lewalter[1]
340	1.59	7	140 (114–186)	Lewalter[1]
80	2.74	12	250 (201–264)	Lewalter[1]
35	3.41	15	280 (250–315)	Lewalter[1]
./.	5.00	22	450*	

* extrapolation
[1] personal communication to the Commission

Despite extensive case reports of occupational exposure to cresols and studies with significant correlations between the external exposure to cresols and the excretion of cresols in urine, there are to date no studies of the quantitative relationship between exposure to cresols and effect parameters. In view of what is known from exposure to phenol and intoxication with cresols (see Table 1), it can, however, be assumed that the systemic toxic effects of cresol are caused by free cresols (Lewalter and Neumann 1998).

Acute intoxication with cresols, like intoxication with phenol, in some cases caused significantly increased serum creatinine values (see Table 1).

Comparison of the results of the biological monitoring of persons with acute intoxication with cresols showed that even after intensive decontamination of the skin, the serum creatinine level can increase significantly above the normal value of 1.0 (0.55–1.1) mg/dl when, with total cresol levels excreted in the urine of up to 200 mg/l, the proportion of free cresols detected in the urine exceeds a "threshold value" of 150 µg/l (Lewalter, personal communication to the Commission).

Serum creatinine values can be used in occupational medicine for monitoring the course of the intoxication, but are not suitable, however, as effect parameters in accordance with the BAT value concept (Lewalter *et al.* 1993).

4 Selection of the Indicators

The exposure and effect indicators available are not specific, at least for *o*-cresol and *p*-cresol, as a result of the ubiquitous background exposure to these substances. Internal exposure to cresols is best evaluated by determining the amounts of free cresols and total cresols (free cresols and conjugated cresols) excreted in urine.

5 Methods

Differentiation of the excreted cresols according to the free and conjugated forms can be carried out on the basis of hydrolysis with perchloric acid (Baselt 1980, De Smet et al. 1998, Teisinger et al. 1954) or enzymatically with β-D-glucuronidase (Piotrowski 1971). The often recommended sulfuric acid hydrolyses only some of the cresyl glucuronide (Piotrowski 1971). After cleaning by means of water vapour distillation or with solid-phase cartridges, the cresol is quantified using high-performance liquid chromatography. A reliable and practicable method has been published by the working group "Analyses of Hazardous Substances in Biological Materials" of the DFG Commission for the Investigation of Health Hazards of Chemical Compounds in the Work Area (Angerer and Schaller 1988).

The direct determination of cresyl sulfate and cresyl glucuronide using high-performance liquid chromatography without hydrolysis and concentration also described, is, despite lower sensitivity, also suitable for the determination of the isomeric cresols in urine (Ogata and Yamasaki 1979). The creatinine values are determined enzymatically (Thomas 2000).

6 Background Exposure

o-Cresol and *m*-cresol do not occur in physiological metabolism, *p*-cresol is formed during the bacterial breakdown of amino acids in the intestine (see Section 1.2).

The background exposure to *o*-cresol in urine can be up to 200 µg/g creatinine and is mainly dependent on smoking habits. In non-smokers and the general population not exposed to the substance, *o*-cresol concentrations of about 60 µg/g creatinine were observed, probably as a result of the ubiquitous exposure to toluene. Smokers, however, were found to have concentrations which were three to four times higher (Nise 1992).

The reference value for the physiological excretion of *p*-cresol in urine is 70 mg/g creatinine (Lewalter et al. 1993).

7 Evaluation of the BLW Value

In general, inhaled and percutaneously absorbed cresols are rapidly detoxified by glucuronidation. The critical toxic effects of cresol metabolism in the kidneys are caused, however, mainly by free cresols. Normally no free cresols can be detected in plasma, even after exposure to high cresol concentrations.

However, as a result of the polymorphism of the isoenzymes of uridine diphosphate glucuronosyl transferase (UDPG), around 6 % of Central Europeans can glucuronidize only slowly. In rare cases this defect can be amplified by the lack of phosphoadenosine-phosphosulfate (PAPS) sulfotransferase normally responsible for sulfation (Calabrese 1986, Lewalter and Neumann 1998). In these persons high exposure to cresols leads to the increased formation of free cresols, which within a few hours can cause massive renal functional disorders and even acute kidney failure (Lewalter and Neumann 1998). The nephrotoxic effects of exposure to high cresol concentrations, which are evidently based on deficient glucuronidation, can be prevented by limiting the amount of free cresol in the total amount of cresol excreted in urine (Lewalter and Neumann 1998, Lewalter et al. 1993, Roots et al. 1992).

Increases in the serum creatinine level of over 1 mg/dl are frequently observed when the amount of free cresol detectable in urine exceeds 150 µg/l (Lewalter and Neumann 1998). A threshold value based on free cresols that could be used as a BAT value, has to date not yet been adequately validated. In the light of what is known from industrial contact with the more toxic phenols (see the documentation on "Phenol" in Volume 1 of this series; Henschler and Lehnert 1994), it can be assumed that also for the cresols a value of 150 µg free cresols/l urine in general will not be exceeded by 95 % of persons exposed to cresols with total cresol levels of up to 200 mg/l urine. In view of this, a BLW value has been set of

200 mg total cresol/l urine

Sampling should be carried out at the end of exposure or at the end of the shift.

For 5 % of persons exposed to cresol (UDPG-deficient persons), the determination of total cresol alone is, however, not sufficient. In such cases, additional determination of free cresol is recommended. The BLW value for total cresol is, therefore, to be regarded as provisional.

8 Interpretation of Data

The BLW value does not take into account local effects resulting from the direct contact with cresols of the skin. Even cresol corrosion on small areas of skin should be reported to the physician.

The problem resulting from the excretion of cresol without occupational exposure to cresols or toluene was discussed in the second addendum to the BAT documentation for toluene from 2000.

9 References

ACGIH (American Conference of Governmental Industrial Hygienists) (2001) Cresol. All isomers. TLV® Chemical substances 7th edition, Documentation. Publication # 7DOC-146, ACGIH®, Cincinatti

Angerer J, Schaller KH (1988) Phenole und Aromatische Alkohole. in: Henschler D (Ed.) Analytische Methoden zur Prüfung gesundheitsschädlicher Arbeitsstoffe. Band 2: Analysen in biologischem Material, 9. Lieferung, VCH, Weinheim

ATSDR (Agency for Toxic Substances and Disease Registry) (1992) Toxicological profile for cresols. US Department of Health and Human Services, USA

Baselt RG (1980) Urine Phenol. in: Baselt RC (Ed.) Biological monitoring methods for industrial chemicals, Biomedical Publications, Davis, 41–42

Bieniek G (1997) Urinary excretion of phenols as an indicator of occupational exposure in the coke-plant industry. Int Arch Occup Environ Health 70: 334–340

Calabrese EJ (1986) Ecogenetics: historical foundation and current status. J Occup Med 28: 1096–1102

Chan TK, Mak LW, Ng RP (1971) Methemoglobinemia, Heinz bodies and acute massive intravascular hemolysis in lysol poisoning. Blood 38: 739–744

DECOS (Dutch Expert Committee on Occupational Standards) (1998) Cresols. Health-based recommended occupational exposure limit. Report of the Dutch expert committee on occupational standards, a committee of the Health Council of the Netherlands, to the Minister and State Secretary of Social Affairs and Employment. No. 1998/15WGD, Rijswijk

Deichmann WB, Keplinger ML (1981) Phenols and phenolic compounds. in: Clayton GD, Clayton FE (Eds) Patty's industrial hygiene and toxicology, 3rd revised edition, Vol. 2A Toxicology, John Wiley and Sons, New York–Chichester–Brisbane–Toronto–Singapore, 2567–2627

De Smet R, David F, Sandra P, Van Kaer J, Lesaffer G, Dhondt A, Lameire N, Vanholder R (1998) A sensitive HPLC method for the quantification of free and total *p*-cresol in patients with chronic renal failure. Clin Chim Acta 278: 1–21

Greim H (Ed.) (2000a) Cresols (all isomers). in: Occupational toxicants, Vol. 14, Wiley-VCH, Weinheim

Greim H (Ed.) (2000b) Kresol (alle Isomeren). Gesundheitsschädliche Arbeitsstoffe, Toxikologisch-arbeitsmedizinische Begründungen von MAK-Werten, 30. Lieferung, Wiley-VCH, Weinheim

Henschler D, Lehnert G (Eds) (1994) Phenol. in: Biological exposure values for occupational toxicants and carcinogens – Critical data evaluation for BAT and EKA values, Vol. 1, VCH, Weinheim, pp. 123–128

Lewalter J, Neumann H-G (1998) Biologische Arbeitsstoff-Toleranzwerte (Biomonitoring). Teil XII: Die Bedeutung der individuellen Empfindlichkeit beim Biomonitoring. Arbeitsmed Sozialmed Umweltmed 33: 352–364

Lewalter J, Steffens W, Wimber M (1993) Biomarker der Polymorphismen im Benzol- und Phenol-Stoffwechsel. in: Triebig G, Stelzer O (Eds) Dokumentationsband über die 33. Jahrestagung der Deutschen Gesellschaft für Arbeitsmedizin und Umweltmedizin e.V., Gentner-Verlag, Stuttgart, 209–214

Moeschlin S (1972) Klinik und Therapie der Vergiftungen, G. Thieme Verlag, Stuttgart, 274–275

Nise G (1992) Urinary excretion of *o*-cresol and hippuric acid after toluene exposure in rotogravure printing. Int Arch Occup Environ Health 63: 377–381

Ogata M, Yamasaki Y (1979) High performance liquid chromatography for the quantitative determination of urinary phenylsulfate and phenylglucuronide as indices of benzene and phenol exposure in rats. Int Arch Occup Environ Health 44: 177–184

Piotrowski JK (1971) Evaluation of exposure to phenol: absorption of phenol vapour in the lungs and through the skin and excretion of phenol in urine. Br J Ind Med 28: 172–178

Roots I, Drakoulis N, Brockmöller J (1992) Polymorphic enzymes and cancer risk: concepts, methodology and data review. in: Kalow W (Ed) Pharmacogenetics of drug metabolism, Pergamon Press, New York, 815–841

Sherwood RJ, Carter FWG (1970) The measurement of occupational exposure to benzene vapour. Am Occup Hyg 13: 125–146

Teisinger J, Fiserova-Bergerova V, Kudra J (1954) The metabolism of benzene in man. Prac Lek Ces 82: 175–188

Thompson DC, Perera K, London R (1996) Studies on the mechanism of hepatotoxicity of 4-methylphenol (*p*-cresol) effects of deuterium labeling and ring substitution. Chem Biol Interact 101: 1–11

Thomas L (2000) Creatinin. in: Thomas L (Ed.) Labor und Diagnose, 5. Auflage, TH-Books Verlagsgesellschaft, Frankfurt/Main, 376–384

WHO (World Health Organization) (1995) Cresols. Environmental Health Criteria 168, Geneva

Authors: J. Lewalter, A.W. Rettenmeier, G. Leng
Approved by the Working Group: 05.05.2003

Methyl bromide

BLW	12 mg bromide/l plasma or serum Sampling time: after several previous shifts
Date of evaluation	2002
Synonyms	Bromomethane Monobromomethane
Formula	CH_3Br
CAS No.	74-83-9
Molecular weight	94.94
Melting point	−93.66°C
Boiling point	3.56°C
Vapour pressure at 15°C	53 kPa
MAK [last established: 1992]	Carcinogen category 3B

(See also the documentation for carcinogenic substances)

At room temperature methyl bromide is a colourless gas, and in low concentrations odourless. It is used as a reaction gas in ionization chambers, for defatting wool and for extracting fats and ethereal oils from nuts, seeds and flowers. It is mainly used, however, as a fumigant in the production, improvement, storage and transport of foodstuffs; such fumigation is prescribed for export of the goods to various countries (e.g. the USA) (see GefStoffV 2003, § 15d). Another important application is in the disinfection of soil. Its use as a fire-extinguishing agent is obsolete (Budavari 1989, Hallier 1996).

1 Metabolism and Kinetics

1.1 Absorption and distribution

Methyl bromide is rapidly absorbed by the organism via the lungs (Andersen *et al.* 1980). Absorption of the substance through the skin is also possible (it is designated with an "H" in the List of MAK and BAT Values). In the blood serum of persons who had direct skin contact with methyl bromide, increased bromide levels were found (Hezemans-Boer *et al.* 1988, Longley and Jones 1965), even if the persons wore adequate respiratory protection (Zwaveling *et al.* 1987).

1.2 Metabolism

Investigations with rats and ^{14}C-labelled methyl bromide showed that around half of the absorbed ^{14}C dose is exhaled as $^{14}CO_2$ (Bond et al. 1985). The rest of the radioactivity is found mainly in the urine, and in small amounts also in the faeces; the distribution between the elimination routes depends on the mode of administration (inhalation exposure, oral administration or intraperitoneal administration) (Medinsky et al. 1985). Incorporated methyl bromide is rapidly metabolized; during this process the bromide ion is cleaved from the methyl group (Bond et al. 1985, Medinsky et al. 1985). The metabolic pathways of methyl bromide correspond with those of the chemically related substances methyl chloride and methyl iodide. They are shown in Figure 1 according to an outline from Kornbrust and Bus (1982).

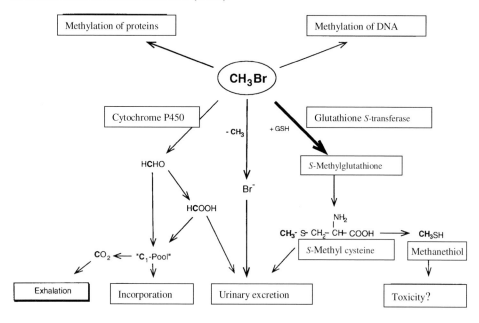

Figure 1. Metabolic pathways of methyl bromide (from Kornbrust and Bus 1982)

A small amount of the monohalogen methanes is oxidized by the cytochrome-P-450 system with cleavage of the halogen ion (here: Br⁻) to form formaldehyde and then formic acid (Kornbrust and Bus 1983).

Methyl halogenides can be enzymatically conjugated in human erythrocytes to form *S*-methyl glutathione (Redford-Ellis and Gowenlock 1971). There are, however, considerable species differences. This metabolic pathway was not detected in the blood of mice, rats, cows, sheep, pigs and rhesus monkeys (Deutschmann et al. 1989).

If human blood samples from different persons are incubated with methyl halogenides, these are conjugated in most cases with *L*-glutathione ("conjugators"), while some are not able to do this ("non-conjugators") (Hallier et al. 1990a, 1990b, Müller et al. 2001a, Thier et al. 1998). The enzyme involved is glutathione *S*-transferase T1 (GSTT1) (Pemble et al. 1994).

In addition to methyl bromide, also methyl chloride and methyl iodide (Hallier et al. 1990a, 1990b), ethylene oxide (Gansewendt et al. 1991) and dichloromethane (Thier et al. 1991) are subject to the same genetic polymorphic metabolism in the erythrocytes. For all the substrates named, GSTT1 enzyme activity is found in the blood of "conjugators", but not in the blood of "non-conjugators". The "conjugators" can, in addition, be divided into individuals with moderate and those with very high enzyme activity, so-called "normal conjugators" and "super conjugators" (Müller et al. 2001b, Thier et al. 1998).

With methyl bromide, GSTT1 activity is much higher than with the other substrates (Schröder et al. 1992).

2 Critical Toxicity

2.1 Acute toxicity

Methyl bromide is an effective pest control agent as a result of its high acute toxicity. This is the reason for the strongly toxic effects in man, which by 1983 had caused more than 950 deaths (Alexeeff and Kilgore 1983).

Above all, methyl bromide damages the central nervous system (Schuler 1998). Depending on the dose, this produces symptoms which range from headaches, visual disorders and vomiting to somnolence and even acute psychosis (Zatuchni and Hong 1981), severe tonic clonic seizures (Behrens and Dukes 1986, Hustinx et al. 1993), paralysis and coma (Kantarjian and Shaheen 1963). The damage to the CNS can lead via severe cerebral oedema to death, in particular as a result of respiratory arrest (Marraccini et al. 1983). If the patients survive, these symptoms usually regress, but permanent effects have been described.

This neurotoxicity can be modulated by GSTT1 polymorphism, as shown in several case reports (Buchwald and Müller 2001, Garnier et al. 1996).

In addition to damage to the nervous system, toxic pulmonary oedema can develop, which can also lead to death (Hine 1969). Other acute effects are kidney damage (tubular necrosis), which can even lead to uraemia, and also impairments in liver function and in blood coagulation with haemorrhage (Prain and Smith 1952). With low and moderate intoxication, the symptoms of intoxication develop after a latency period of several hours (Hine 1969). Methyl bromide causes reddening of the skin and mucous membranes, which leads to swelling and blistering and even severe ulcers and necrosis (Hezemans-Boer et al. 1988). After skin contact with the substance, at first no symptoms of irritation are visible; the damage becomes evident after a latency period of at least an hour (Longley and Jones 1965, Zwaveling et al. 1987).

2.2 Chronic toxicity

The long-term effects of methyl bromide, apart from the symptoms remaining after severe acute intoxication, are less pronounced. Changes were described which become manifest after acute intoxication occurs on top of chronic subclinical intoxication (Alexeeff and Kilgore 1983). Neurological symptoms were the main finding in the comparison of occupationally exposed persons and control persons; in particular EEG changes (Mellerio *et al.* 1974, Verberk *et al.* 1979) and poorer performance of the exposed persons in neuropsychological performance tests were reported (Anger *et al.* 1986). In addition, the transaminases in serum may be increased (Tanaka *et al.* 1991).

Methyl bromide is mutagenic in various test systems (Djalali-Behzad *et al.* 1981, Voogd *et al.* 1982). In human lymphocytes it induces *in vitro* chromosomal aberrations and increases sister chromatid exchange (SCE) (Garry *et al.* 1990, Tucker *et al.* 1985). The GSTT1 polymorphism already described had a modulating influence also on sister chromatid exchange. On the basis of blood samples from various persons it could be shown *in vitro* that "conjugators" (who react more sensitively to the neurotoxic effects of methyl bromide than "non-conjugators") react less sensitively to cytogenetic toxicity than "non-conjugators" (Hallier *et al.* 1993). As methyl bromide was found to be genotoxic *in vitro* and *in vivo* and part of the human population is particularly sensitive to the genotoxic effects of methyl bromide (Bolt and Gansewendt 1993), the substance was classified in the former section IIIB (Greim 1996) (corresponding to the current category 3B). A comprehensive evaluation of the carcinogenic potential of methyl bromide was published by the International Agency for Research on Cancer (IARC 1999).

3 Exposure and Effects

3.1 External exposure

As occupational-medical field studies have shown, the meaningfulness of the results in the case of methyl bromide fumigators is relatively small, as inhalation of the methyl bromide is more or less prevented by respiratory protection and under such conditions of application the substance mainly enters the body through the skin (Guillemin *et al.* 1990, van den Oever *et al.* 1984). In addition, the exposure profile, with very short exposure peaks and a strong dependence on the spatial conditions of the object to be fumigated and current climatic conditions, varies considerably.

3.2 Internal exposure

To date, methyl bromide cannot be detected in the organism with the usual analytical methods. The internal exposure can be determined on the basis of the following metabolites: bromide in urine or serum, or S-methyl cysteine in urine.

The *biologically effective dose* can be determined in the form of DNA or protein adducts in blood. These parameters are an improvement on the determination of the substance itself or a metabolite as they show not only the exposure but also the effects of it.

Early biological effects are changes in the germ plasm. Parameters that can be used are e.g. the detection of DNA strand breaks and cross-links, increased DNA repair (UDS), micronuclei, sister chromatid exchange and chromosomal aberrations.

4 Selection of the Indicators

As a result of occupational-medical field studies during fumigation with methyl bromide the various indicators can be evaluated as follows:

1. The detection of unchanged *methyl bromide in blood* is not possible at present with routine methods as a result of the short life-span of the molecule in the organism (Hustinx et al. 1993). It is also unsuitable as a parameter for biological monitoring as the blood must be sampled during the exposure.
2. The *bromide concentration in blood or urine* is suitable as a parameter for internal exposure. Bromide is released from methyl bromide both as a result of the reaction with *L*-glutathione and the alkylation of macromolecules (see Figure 1). The levels of bromide in urine correlate with the external exposure to methyl bromide (Tanaka et al. 1991). Bromide has a long half-life in human blood of 12–14 days (Söremark 1960, Vaiseman et al. 1986) and is therefore also a good cumulative dosimeter for the determination of internal exposure after long-term exposure to methyl bromide. The bromide levels in serum and plasma therefore allow the clear differentiation of persons exposed to methyl bromide and collectives not exposed (Müller et al. 1999, 2001c). Bromide values of 12 mg/l blood and above, determined after long-term fumigation during the fumigation season, are accompanied by slight changes in the EEG of the fumigators (Verberk et al. 1979). Depending on the severity of the symptoms, which ranged from drowsiness, nausea and vomiting to myoclonus and generalized clonic tonic convulsions, nine persons intoxicated with methyl bromide were found to have bromide levels between 51 and 363 mg/l serum (determined 12 hours after exposure) (Hustinx et al. 1993).
3. The *excretion of mercapturic acid in urine* is suitable for the monitoring of exposure to substances which are metabolized mainly via conjugation with *L*-glutathione. Earlier studies showed, however, that in man the amount of mercapturic acids excreted is small after exposure to methyl bromide (Thompson et al. 1966). An explanation for this is provided by animal studies of Kornbrust and Bus (1983, 1984) and Medinsky et al. (1985): methyl halogenides are metabolized via *S*-methyl

glutathione to *S*-methyl cysteine and methyl mercaptan, which is held responsible for the neurotoxic effects.
4. The *excretion of S-methyl cysteine in the urine* of persons exposed to methyl bromide can be detected (Deutschmann *et al.* 1990). This parameter could be suitable for the monitoring of fumigators.
5. The *DNA strand breaks in lymphocytes* are a short-term parameter for the exposure of the genetic material. The complicated analytical procedure and the influence of non-occupational factors make the application of this method difficult. The determination of the DNA strand breaks in lymphocytes can, however, be suitable for certain scientific questions.
6. Cytogenetic effects, such as *sister chromatid exchange* in lymphocytes, are suitable for investigating certain questions. The time and personnel needed for SCE determination, however, make the procedure impractical for routine practice. The main disadvantage of this method is the great interindividual variability.
7. The determination of *reaction products (adducts)* with macromolecules in blood, in particular serum albumin and haemoglobin, was found to be a suitable parameter for the biological monitoring of exposure to methyl bromide during fumigation. *S*-methyl cysteine in haemoglobin is the parameter suggested by various working groups (Bailey *et al.* 1981, Iwasaki 1988). Methyl bromide has a clear preference for cysteine residues (Ferranti *et al.* 1996), in particular for the cysteine at position 93 of β-globin (Sannolo *et al.* 1999). *N*-terminal valine, however, is only of subordinate importance in the alkylation reaction of methyl bromide. The use of *S*-methyl cysteine in serum albumin has also been described as a biological monitoring parameter (Müller *et al.* 1994). An advantage of the determination of *S*-methyl cysteine in haemoglobin and in serum albumin is that this parameter is not influenced by smoking habits (Hallier 1996, Iwasaki *et al.* 1989, Sannolo *et al.* 1999). The considerable interindividual variability of the adduct levels in haemoglobin (Iwasaki *et al.* 1989) and serum albumin (Garnier *et al.* 1996) observed in fumigators exposed to methyl bromide is the result of the modulation by the polymorphic GSTT1 metabolism of the methyl bromide concentrations available for alkylation (Müller *et al.* 1998). "Non-conjugators" therefore seem to have higher protein adduct levels (Garnier *et al.* 1996).

5 Methods

Bromide in plasma/serum can be determined with great sensitivity (detection limit 1 mg/l) using a simple, newly developed, standard photometric procedure (Müller *et al.* 1999). The procedure was tested and approved by the working group "Analyses of Hazardous Substances in Biological Materials" of the DFG Commission for the Investigation of Health Hazards of Chemical Compounds in the Work Area (Angerer and Schaller 2005).

The analysis of *S*-methylated albumin and globin from human blood was described in detail (Müller *et al.* 1994). It includes protein isolation and hydrolysis, fluorescence derivatization, and HPLC separation and fluorescence detection of the *S*-methyl cysteine.

The determination of the GSTT1 conjugator status (phenotyping) can be carried out by measuring the substrate reduction kinetics using a standard headspace-GC procedure (Peter *et al.* 1989), which has been tested and published (Angerer and Müller 2004). A newly developed method is based on the use of HPLC with fluorescence detection (Müller *et al.* 2001a, 2001b).

6 Background Exposure

The bromide ion, like the chloride ion, is distributed mainly in the extracellular compartments (Rauws 1983). In all body fluids such as whole blood, serum/plasma and urine, bromide concentrations are found in the range from 3–12 mg/l with a mean value of around 5 mg/l (Olszowy *et al.* 1998). Investigations of the background exposure to bromide in whole blood of 183 randomly selected blood donors in Australia yielded a mean value of 5.3 ± 1.4 mg/l (Olszowy *et al.* 1998). Investigations with plasma from 74 healthy Central Europeans living in Germany yielded a median bromide value of 4.8 mg/l plasma as the background exposure; the arithmetic mean was 4.6 ± 1.2 mg/l plasma. The 95th percentile of the individual values determined was 6.4 mg/l plasma in this study (Müller *et al.* 2001c). While the above values correlate well with each other, in Chilean collectives median bromide values for background exposure of 12.6 mg/l plasma (blood donors of an urban population) and 11.1 mg/l plasma (agricultural workers not occupationally exposed to bromide compounds) were found (Müller *et al.* 2001c). In the same study, the median plasma bromide levels of persons exposed to methyl bromide were 15.4 mg/l plasma and thus clearly higher than those for background exposure in the collectives of not exposed Chileans. These investigations show that a uniform international value for background exposure to bromide in plasma cannot be given, while a regional value for background exposure is possible, which may depend among other things on the eating habits of the local population.

For the protein adducts the base value for the general population is 15 nmol *S*-methyl cysteine per g albumin or haemoglobin (corresponding to 2 µg/g albumin or haemoglobin, and 42 or 315 µg/l blood) (Bailey *et al.* 1981, Müller *et al.* 1994). The 95th percentile of all individual values was 18.2 nmol *S*-methyl cysteine/g albumin (corresponding to 2.5 µg/g albumin) (Hallier 1996).

7 Evaluation of the BLW Value

The high acute and chronic neurotoxicity (Anger *et al.* 1986, Calvert *et al.* 1998) of methyl bromide makes adequate monitoring of the exposure of the employees imperative. Methyl bromide can pass through the skin and as a rule the users wear respiratory protection. Under these circumstances effective exposure control is only possible by means of biological monitoring.

On the basis of the available case reports of intoxication with methyl bromide (e.g. De Haro *et al.* 1997) and the results of Verberk *et al.* (1979), the occurrence of neurotoxic effects relevant to occupational medicine is to be expected at bromide concentrations above 20 mg/l plasma or serum. Below 20 mg/l plasma slight neurotoxic effects cannot be excluded. The background exposure of the German population is given as 6.5 mg/l plasma as a reference value (95th percentile) (Müller *et al.* 1999, Olszowy *et al.* 1998).

In view of this, a BLW value has been set of

12 mg bromide/l plasma or serum

8 Interpretation of Data

As a result of the long half-life of bromide in human blood of 12–14 days (Söremark 1960, Vaiseman *et al.* 1986) the parameter is well suited for use as a cumulative dosimeter for determining long-term exposure to methyl bromide. Sampling should be carried out after several previous shifts.

The background exposure to bromide in plasma is mainly the result of the unavoidable uptake of the anion from food and drinking water (Olszowy *et al.* 1998). Increased bromide levels can be the result of occupational and environmental exposure to bromide, bromine and substances and medicines containing bromine. To determine a specific exposure to bromide or a substance that releases bromide, it is necessary to exclude exposure of the person to other sources of bromide by means of a thorough occupational-medical or environmental-medical anamnesis. Attention is drawn in particular to medicines containing bromide and the drugs dextromethorphan bromide, a component of numerous antitussives, and pyridostigmine bromide, which is used as a protective agent for inhibition of acetylcholinesterase (Szinicz 1994).

9 References

Alexeeff GV, Kilgore WW (1983) Ethyl bromide. Residue Rev 88: 101–153
Andersen ME, Gargas ML, Jones RA, Jenkins LJ Jr (1980) Determination of the kinetic constants for metabolism of inhaled toxicants *in vivo* using gas uptake measurement. Toxicol Appl Pharmacol 54: 100–116
Anger WK, Moody L, Burg J, Brightwell WS, Taylor BJ, Russo JM, Dickerson N, Setzer JV, Johnson BL, Hicks K (1986) Neurobehavioral evaluation of soil and structural fumigators using methyl bromide and sulfuryl fluoride. Neurotoxicology 7: 137–156
Angerer J, Müller M (2004) Glutathione *S*-transferase T1 (GSTT1) phenotyping. in: Angerer J, Müller M (Eds) Analyses of hazardous substances in biological materials, Vol. 9, Special issue: Markers of susceptibility, Wiley-VCH, Weinheim

Angerer J, Schaller KH (Eds) (2005) Bromid in Serum. in: Analysen in biologischem Material, Loseblattsammlung, 17. Lieferung, Wiley-VCH, Weinheim (in press)

Bailey E, Connors TA, Farmer PB, Gorf SM, Rickard J (1981) Methylation of cysteine in hemoglobin following exposure to methylating agents. Cancer Res 41: 2514–2517

Behrens RH, Dukes DC (1986) Fatal methyl bromide poisoning. Br J Ind Med 43: 561–562

Bolt HM, Gansewendt B (1993) Mechanisms of carcinogenicity of methyl halides. Crit Rev Toxicol 23: 237–253

Bond JA, Dutcher JS, Medinsky MA, Henderson RF, Birnbaum LS (1985) Disposition of [^{14}C]methyl bromide in rats after inhalation. Toxicol Appl Pharmacol 78: 259–267

Buchwald AL, Müller M (2001) Late confirmation of acute methyl bromide poisoning using S-methylcysteine adduct testing. Vet Hum Toxicol 43: 208–211

Budavari S (Ed.) (1989) The Merck Index – Encyclopaedia of chemicals, drugs, and biologicals, 11th edition, Merck & Co. Inc., Rahway NJ, USA, 834

Calvert GM, Mueller CA, Fajen JM, Chrislip DW, Russo J, Briggle T, Fleming LE, Suruda AJ, Steenland K (1998) Health effects associated with sulfuryl fluoride and methyl bromide exposure among structural fumigation workers. Am J Public Health 88: 1774–1780

De Haro L, Gastaut JL, Jouglard J, Renacco E (1997) Central and peripheral neurotoxic effects of chronic methyl bromide intoxication. J Toxicol Clin Toxicol 35: 29–34

Deutschmann S, Peter H, Reichel C, Bolt M, Hallier E (1989) Kinetik der Metabolisierung von Methylbromid und Methyliodid in Erythrocyten des Menschen und verschiedener Tierspezies. in: Meyer-Falcke A, Jansen G (Eds) Dokumentationsband über die 29. Jahrestagung der Deutschen Gesellschaft für Arbeitsmedizin und Umweltmedizin e.V., Gentner-Verlag, Stuttgart, 517–519

Deutschmann S, Peter H, Schröder K, Goch S, Hallier E (1990) Bestimmung von S-Methylglutathion in Erythrocyten und von S-Methylcystein im Urin als Indikatoren einer Belastung mit Methylbromid. in: Schuckmann F, Schopper-Jochum S (Eds) Dokumentationsband über die 30. Jahrestagung der Deutschen Gesellschaft für Arbeitsmedizin und Umweltmedizin e.V., Gentner-Verlag, Stuttgart, 509–511

Djalali-Behzad G, Hussain S, Osterman-Golkar S, Segerbäck D (1981) Estimation of genetic risks of alkylating agents. VI: Exposure of mice and bacteria to methyl bromide. Mutat Res 84: 1–9

Ferranti P, Sannolo N, Mamone G, Fiume I, Carbone V, Tornqvist M, Bergman A, Malorni A (1996) Structural characterization by mass spectrometry of hemoglobin adducts formed after *in vivo* exposure to methyl bromide. Carcinogenesis 17: 2661–2671

Gansewendt B, Föst U, Xu D, Hallier E, Bolt HM, Peter H (1991) Formation of DNA adducts in F-344 rats after oral administration or inhalation of [^{14}C]methyl bromide. Food Chem Toxicol 29: 557–563

Garnier R, Rambourg-Schepens MO, Müller A, Hallier E (1996) Glutathione transferase activity and formation of macromolecular adducts in two cases of acute methyl bromide poisoning. Occup Environ Med 53: 211–215

Garry VF, Nelson RL, Griffith J, Harkins M (1990) Preparation for human study of pesticide applications: sister chromatid exchanges and chromosome aberrations in cultured human lymphocytes exposed to selected fumigants. Teratog Carcinog Mutagen 10: 21–29

GefStoffV (Gefahrstoffverordnung) (2003) Edited: Weinmann W, Thomas HP, Klein A, C. Heymanns Verlag, Köln–Berlin–Bonn–München

Greim H (Ed.) (1996) Methyl bromide. in: Occupational toxicants, Vol. 7, Wiley-VCH, Weinheim

Guillemin MP, Hillier RS, Bernhard CA (1990) Occupational and environmental hygiene assessment of fumigations with methyl bromide. Ann Occup Hyg 34: 591–607

Hallier E, Deutschmann S, Reichel C, Bolt HM, Peter H (1990a) A comparative investigation of the metabolism of methyl bromide and methyl iodide in human erythrocytes. Int Arch Occup Environ Health 62: 221–225

Hallier E, Jaeger R, Deutschmann S, Bolt HM, Peter H (1990b) Glutathione conjugation and cytochrome P-450 metabolism of methyl chloride *in vitro*. Toxicology *in vitro* 4: 513–517

Hallier E, Langhof T, Dannappel D, Leutbecher M, Schröder K, Goergens HW, Müller A, Bolt HM (1993) Polymorphism of glutathione conjugation of methyl bromide, ethylene oxide and dichloromethane in human blood: influence on the induction of sister chromatid exchanges (SCE) in lymphocytes. Arch Toxicol 67: 173–178

Hallier E (1996) Arbeitsmedizinische Untersuchungen zur Problematik der Durchführung von Begasungen mit Methylbromid. Habilitationsschrift, Ruhr-Universität Bochum. Deutsche Hochschulschriften 1089, Verlag Hänsel-Hohenhausen, Egelsbach–Frankfurt–St. Peter Post

Hezemans-Boer M, Toonstra J, Meulenbelt J, Zwaveling JH, Sangster B, van Vloten WS (1988) Skin lesions due to exposure to methyl bromide. Arch Dermatol 124: 917–921

Hine CH (1969) Methyl bromide poisoning. A review of ten cases. J Occup Med 11: 1–10

Hustinx WNM, van de Laar RTH, van Huffelen AC, Verwey JC, Meulenbelt J, Savelkoul TJF (1993) Systemic effects of inhalational methyl bromide poisoning: a study of nine cases occupationally exposed due to inadvertent spread during fumigation. Br J Ind Med 50: 155–159

IARC (International Agency for Research on Cancer) (1999) Methyl bromide. IARC Monogr Eval Carcinog Risks Hum 71: 721–735

Iwasaki K (1988) Determination of S-methylcysteine in mouse hemoglobin following exposure to methyl bromide. Ind Health 26: 187–190

Iwasaki K, Ito I, Kagawa J (1989) Biological exposure monitoring of methyl bromide workers by determination of hemoglobin adducts. Ind Health 27: 181–183

Kantarjian AD, Shaheen AS (1963) Methyl bromide poisoning with nervous system manifestations resembling polyneuropathy. Neurology 13: 1054–1058

Kornbrust DJ, Bus JS (1982) Metabolism of methyl chloride to formate in rats. Toxicol Appl Pharmacol 65: 135–143

Kornbrust DJ, Bus JS (1983) The role of glutathione and cytochrome P-450 in the metabolism of methyl chloride. Toxicol Appl Pharmacol 67: 246–256

Kornbrust DJ, Bus JS (1984) Glutathione depletion by methyl chloride and association with lipid peroxidation in mice and rats. Toxicol Appl Pharmacol 72: 388–399

Longley EO, Jones AT (1965) Methyl bromide poisoning in man. Ind Med Surg 34: 499–502

Marraccini JV, Thomas GE, Ongley JP, Pfaffenberger CD, Davis JH, Bednarczyk LR (1983) Death and injury caused by methyl bromide, an insecticide fumigant. J Forensic Sci 28: 601–607

Medinsky MA, Dutcher JS, Bond JA, Henderson RF, Mauderly JL, Snipes MB, Mewhinney JA, Cheng YS, Birnbaum LS (1985) Uptake and excretion of [^{14}C]methyl bromide as influenced by exposure concentration. Toxicol Appl Pharmacol 78: 215–225

Mellerio F, Gaultier M, Bismut C (1974) Electroencephalography during acute poisoning by methyl bromide. Eur J Toxicol Environ Hyg 7: 119–132

Müller AMF, Hallier E, Westphal G, Schröder KR, Bolt HM (1994) Determination of methylated globin and albumin for biomonitoring of exposure to methylating agents using HPLC with precolumn fluorescent derivatization. Fresenius J Anal Chem 350: 712–715

Müller M, Krämer A, Angerer J, Hallier E (1998) Ethylene oxide protein adduct formation in humans: influence of glutathione-S-transferase polymorphisms. Int Arch Occup Environ Health 71: 499–502

Müller M, Reinhold P, Lange M, Zeise M, Jürgens U, Hallier E (1999) Photometric determination of human serum bromide levels – a convenient biomonitoring parameter for methyl bromide exposure. Toxicol Lett 107: 155–159

Müller M, Voss M, Heise C, Schulz T, Bünger J, Hallier E (2001a) High-performance liquid chromatography/fluorescence detection of S-methylglutathione formed by glutathione-S-transferase T1 *in vitro*. Arch Toxicol 74: 760–767

Müller M, Voss M, Schulz T, Hallier E (2001b) Eine Routine-HPLC-Fluoreszenz-Methode für den Suszeptibilitätsparameter humane erythrozytäre Glutathion-S-Transferase T1 Aktivität. Umweltmed Forsch Prax 6: 47–50

Müller M, Barraza X, Reinhold P, Westphal G, Bünger J, Zeise M, Hallier E (2001c) Bestimmung des Plasmabromidspiegels als arbeits- und umweltmedizinischer Belastungsparameter. in: Drexler H, Broding HC (Eds) Dokumentationsband über die 41. Jahrestagung der Deutschen Gesellschaft für Arbeitsmedizin und Umweltmedizin e.V., 319–322, Rindt-Druck, Fulda

van den Oever R, Jacques P, Roosels D, Lahaye D (1984) Health hazards of soil disinfection by methyl bromide fumigation in Belgian greenhouses. Cah Med Trav 21: 211–215

Olszowy HA, Rossiter J, Hegarty J, Geoghegan P (1998) Background levels of bromide in human blood. J Anal Toxicol 22: 225–230

Pemble S, Schroeder KR, Taylor JB, Spencer S, Meyer DJ, Hallier E, Bolt HM, Ketterer B (1994) Human glutathione S-transferase theta (GSTT1): cDNA cloning and the characterization of a genetic polymorphism. Biochem J 300: 271–276

Peter H, Deutschmann S, Reichel C, Hallier E (1989) Metabolism of methyl chloride by human erythrocytes. Arch Toxicol 63: 351–355

Prain JH, Smith GH (1952) A clinical-pathological report of eight cases of methyl bromide poisoning. Br J Med 9: 44–49

Rauws AG (1983) Pharmacokinetics of bromide ion – an overview. Food Chem Toxicol 21: 379–382

Redford-Ellis M, Gowenlock AH (1971) Studies on the reaction of chloromethane with human blood. Acta Pharmacol Toxicol 30: 36–48

Sannolo N, Mamone G, Ferranti P, Basile A, Malorni A (1999) Biomonitoring of human exposure to methyl bromide by isotope dilution mass spectrometry of peptide adducts. J Mass Spec 34: 1028–1032

Schröder KR, Hallier E, Peter H, Bolt HM (1992) Dissociation of a new glutathione S-transferase activity in human erythrocytes. Biochem Pharmacol 43: 1671–1674

Schuler (1998) Vergiftung durch Brommethyl? Dtsch Wochenschr Öffentl Gesundheitspfl 31: 696–704

Söremark R (1960) The biological half-life of bromide ions in human blood. Acta Physiol Scand 50: 119–123

Szinicz L (1994) Chemische Kampfstoffe. in: Marquardt H, Schäfer SG (Eds) Lehrbuch der Toxikologie, BI-Wissenschaftsverlag, Mannheim, 571–588

Tanaka S, Abuku S, Seki Y, Imamiya S (1991) Evaluation of methyl bromide exposure on the plant quarantine fumigators by environmental and biological monitoring. Ind Health 29: 11–21

Thier R, Föst U, Deutschmann S, Schröder KR, Westphal G, Hallier E, Peter H (1991) Distribution of methylene chloride in human blood. Arch Toxicol 14: 254–258

Thier R, Delbanco EH, Wiebel FA, Hallier E, Bolt HM (1998) Determination of glutathione transferase (GSTT1-1) activities in different tissues based on formation of radioactive metabolites using ^{35}S-glutathione. Arch Toxicol 72: 811–815

Thompson RH (1966) A review of the properties and usage of methyl bromide as a fumigant. J Stored Prod Res 1: 353–376

Tucker JD, Xu J, Stewart J, Ong T (1985) Development of a method to detect volatile genotoxins using sister chromatid exchanges. Environ Mutagen 7, Suppl 3: 48

Vaiseman N, Koren G, Pencharz P (1986) Pharmacokinetics of oral and intravenous bromide in normal volunteers. J Toxicol Clin Toxicol 24: 403–413

Verberk MM, Rooyakkers-Beemster T, de Vlieger M, van Vliet AGM (1979) Bromine in blood, EEG and transaminases in methyl bromide workers. Br J Ind Med 36: 59–62

Voogd CE, Knaap AGAC, van der Heijden CA, Kramers PGN (1982) Genotoxicity of methyl bromide in short-term assay systems. Mutat Res 97: 233, Abstr 130

Zatuchni J, Hong K (1981) Methyl bromide poisoning seen initially as psychosis. Arch Neurol 38: 529–530

Zwaveling JH, de Kort WL, Meulenbelt J, Hezemans-Boer M, vam Vloten WA, Sangster B (1987) Exposure of the skin to methyl bromide: a study of six cases occupationally exposed to high concentrations during fumigation. Hum Toxicol 6: 491–495

Authors: E. Hallier, M. Müller
Approved by the Working Group: 10.02.2002

Phenol, Addendum

BLW	200 mg total phenol/l urine
	Sampling time: end of exposure or end of shift
Date of evaluation	2003

The toxicity of phenol was reviewed in Volume 1 of this series (Henschler and Lehnert 1994). This is a supplement to that chapter.

In 1998 phenol was classified in Carcinogen category 3 of Section III of the *List of MAK and BAT Values* and the MAK value was withdrawn. Decisive were the clastogenic and tumour-promoting effects, the close metabolic association with benzene and hydroquinone, and evidence of carcinogenic effects, which, however, cannot be conclusively evaluated (Greim 1998).

The previous BAT value for phenol (300 mg phenol/l urine) was evaluated on the basis of the relationship observed in three studies between the external exposure to phenol and the phenol excreted in urine. The individually tolerable levels of exposure to phenol were based on the main biochemical parameters of effect for creatinine metabolism in the kidneys found in occupational-medical field studies (see Volume 1). As a result of the withdrawal of the MAK value, the BAT value has also been withdrawn. The data available allow a BLW value, however, to be evaluated.

1 Metabolism and Kinetics

Data for the metabolism and kinetics of phenol can be found in Volume 1 of this series (Henschler and Lehnert 1994) and in the MAK documentation (Greim 1998).

Phenol is a physiological product of metabolism. The physiological excretion of phenol in urine (free and conjugated) reaches normal values of 20 to 35 mg/g creatinine with peak values of up to 81.5 mg/g creatinine. The proportion of unbound (free) phenol is given in more recent studies as less than 1 % (see Volume 1).

Phenol is absorbed rapidly via the lungs, the gastrointestinal tract and the intact skin. Ingested phenol is eliminated almost completely with the 24-hour urine; around 70 % to 80 % of the dose is excreted as phenyl sulfate and 15 % to 25 % as phenyl-β-D-glucuronide (see Volume 1).

Reactive hydroxylation products of phenol such as hydroquinone or catechol, as were detected in small amounts in animal experiments or in earlier studies (see Volume 1 of this series (Henschler and Lehnert 1994) and Greim 1998), could not be detected in a group of 483 workers (Lewalter and Neumann 1998). Such compounds have, to date, only been detected during the biological monitoring of benzene metabolism (Lewalter and Neumann 1998).

2 Exposure and Effects

2.1 Relationship between external and internal exposure

There are data available from occupational-medical studies for the correlation between external and internal exposure to phenol (see Table 1). Since the last BAT documentation from 1994 of this series, the previous correlation data have been supplemented for the low dose range by a study with coking plant workers (Bieniek 1997) and by personal communications to the Commission.

Table 1. Correlation between external and internal exposure to phenol after occupational exposure

Air		Urine	References
Phenol (mean shift value)		Total phenol (mean)	
[ml/m^3]	[mg/m^3]	[mg/l urine]	
0.08	0.3	10	Bieniek 1997
0.8	3	75 [50–91]	Lewalter, personal communication to the Commission
1.3	5	100	Piotrowski 1971
2.1	8	160 [135–189]	Lewalter, personal communication to the Commission
3.9	15	300	Ohtsuji and Ikeda 1972
5	19*	451	Ogata et al. 1986

* former MAK value

2.2 Field studies

In a group of 483 workers exposed to phenol concentrations of up to 5 ml/m^3 (former MAK value), no quinoid compounds could be detected down to a detection limit of 0.1 µg hydroquinone/l urine. In 84 % of these workers exposed to phenol, with total phenol levels excreted of up to 190 mg/g creatinine, the plasma creatinine value was within the reference range of 1.0 ± 0.2 mg/dl. In 5 % of the workers, the plasma creatinine value was markedly increased at 1.5 mg/dl, although their total phenol levels were only 168 mg/g creatinine. In this group, despite the comparatively low total phenol levels, the excretion of free phenol in urine was significantly increased at a level of 280 µg/g creatinine. On the other hand, in 11 % of the workers with a plasma creatinine value of 0.6 mg/dl and a total phenol level of 220 mg/g creatinine only 135 µg free phenol/g creatinine was excreted in the urine. It was discovered during the evaluation of this study that the plasma creatinine level was always significantly above the normal value of 1 mg/dl when more than 150 µg free phenol/g creatinine was excreted in the urine (see Table 2; Lewalter and Neumann 1998).

Table 2. Biological monitoring of a group of 483 workers exposed to phenol, sorted according to the creatinine levels in plasma (95th percentile of the average values)

Group		Plasma	Urine		
Number of persons (N)	Proportion of the group (%)	Creatinine [mg/dl]	Total phenol[1] [mg/g creatinine]	Free phenol [µg/g creatinine]	Hydroquinone [µg/l urine]
55	11	0.6 (± 0.2)	220	135	< 0.1
388	84	1.0 (± 0.2)	190	150	< 0.1
23	5	1.5 (± 0.3)	168	280	< 0.1
483	100				

[1] The values given in the publication were corrected.

3 Evaluation of the BLW Value

Phenol absorbed by inhalation or through the skin is generally rapidly detoxified by glucuronidation. The critical toxic effects of phenol metabolism in the kidneys are caused mainly by free phenol. Normally only traces of free phenol can be detected in plasma, even after exposure to high phenol levels.

However, as a result of polymorphism of the isoenzymes of uridine diphosphate glucuronyl transferase (UDPG), around 6 % of Central Europeans can glucuronidize only slowly. In rare cases this defect can be amplified by the lack of phosphoadenosine-phosphosulfate (PAPS) sulfotransferase responsible for sulfation (Calabrese 1986, Lewalter and Neumann 1998). In these persons high exposure to phenol leads to the increased formation of free phenols, which within a few hours can cause severe renal functional disorders and even acute kidney failure (Lewalter and Neumann 1998).

The nephrotoxic effects of high exposure to phenol, which are evidently the result of deficient glucuronidation, can be prevented by limiting the amount of free phenol in the total phenol excreted in urine.

In normal persons such excretion of free phenol is generally observed only after severe intoxication with phenol (Enkin *et al*. 1966).

Damage to the kidneys is observed when the plasma creatinine levels increase to above 1 mg/dl and the proportion of free phenol detectable in urine exceeds 150 µg/g creatinine (Lewalter and Neumann 1998). A value based on free phenol that could be used as a threshold value in biological material, has, to date, however, not yet been adequately validated. A value of 150 µg free phenol/g creatinine is generally not exceeded by 95 % of persons exposed to phenol with values up to 200 mg total phenol/g creatinine, corresponding to a value of 200 mg total phenol/l urine. In view of this, a BLW value has been set of

200 mg total phenol/l urine

Sampling should be carried out at the end of exposure or the end of the shift.

For 5 % of persons exposed to phenol (UDPG-deficient persons), the determination of total phenol alone is, however, not sufficient protection. In such cases, additional determination of free phenol is recommended. The BLW value for total phenol should therefore be regarded as provisional until there are sufficient data available to be able to evaluate a threshold limit value for free phenol.

4 Interpretation of Data

The BLW value does not take into account local effects resulting from contact of phenol with the skin.

As marked effects of phenol can occur independent of the level of exposure to phenol, to promote occupational health and safety, prophylactic monitoring of the 1 to 2-hour gradient of creatinine in plasma is recommended as urgent primary prevention directly after the end of the shift for all unknown and in particular high levels of exposure to phenol (acute intoxication with phenol) (Lewalter and Mischke 1991).

The correlation data are based on the experience with acute intoxication with phenol. As a result of the prophylactic monitoring of the 1 to 2-hour gradient of creatinine in plasma, possible adverse effects of exposure to phenol are usually recognised early also in UDPG-deficient and PAPS-deficient persons.

5 References

Bieniek G (1997) Urinary excretion of phenols as an indicator of occupational exposure in the coke-plant industry. Int Arch Occup Environ Health 70: 334–340

Calabrese EJ (1986) Ecogenetics: historical foundation and current status. J Occup Med 28: 1096–1102

Enkin M, Treuil J, Becache A, Revol L (1966) Study of phenol excretion in clinical psychiatry. II. Total, free, conjugated, acid, volatile phenols and vanyl-mandelic acid. Encephale 55: 391–451

Greim H (Ed.) (1998) Phenol. Gesundheitsschädliche Arbeitsstoffe, Toxikologisch-arbeitsmedizinische Begründungen von MAK-Werten, 27. Lieferung, VCH, Weinheim

Henschler D, Lehnert G (Eds) (1994) Phenol. in: Biological exposure values for occupational toxicants and carcinogens – Critical data evaluation for BAT and EKA values, Vol. 1, VCH, Weinheim, pp. 123–128

Lewalter J, Mischke LW (1991) Empfehlungen zur arbeitsmedizinischen Prävention expositions- und dispositionsbedingter Arbeitsstoff-Beanspruchungen. in: Schäcke G, Ruppe K, Vogel-Sührig C (Eds) Dokumentationsband über die 31. Jahrestagung der Deutschen Gesellschaft für Arbeitsmedizin und Umweltmedizin e.V., Gentner-Verlag, Stuttgart, 135–139

Lewalter J, Neumann HG (1998) Biologische Arbeitsstoff-Toleranzwerte (Biomonitoring). Teil XII: Die Bedeutung der individuellen Empfindlichkeit beim Biomonitoring. Arbeitsmed Sozialmed Umweltmed 33: 352–364

Ogata M, Yamasaki Y, Kawai T (1986) Significance of urinary phenyl sulfate and phenyl glucuronide as indices of exposure to phenol. Int Arch Occup Environ Health 58: 197–202

Ohtsuji H, Ikeda M (1972) Quantitative relationship between atmospheric phenol vapour and phenol in the urine of workers in Bakelite factories. Br J Ind Med 29: 70–73

Piotrowski JK (1971) Evaluation of exposure to phenol: absorption of phenol vapour in the lungs and through the skin and excretion of phenol in urine. Br J Ind Med 28: 172–178

Authors: J. Lewalter, G. Leng
Approved by the Working Group: 05.05.2003

List of Authors and Date of Compilation

Texts	Authors	Compilation
Antimony and its inorganic compounds	K.H. Schaller	2002
Arsenic and inorganic arsenic compounds	H. Drexler	2002
Beryllium and its inorganic compounds	K.H. Schaller	2002
Cresols (all isomers)	J. Lewalter, A.W. Rettenmeier, G. Leng	2003
Cyclohexane	U. Knecht, K.H. Schaller	2001
1,4-Dichlorobenzene, Addendum		
Dichloromethane, Addendum	H.M. Bolt	2001
Formic acid methyl ester	H. Drexler, G. Csanády	2002
Hexachlorobenzene, Addendum	J. Lewalter, U. Reuter	2002
Lead and its compounds	H.M. Bolt, K.H. Schaller	2000 and 2003
Manganese and its inorganic compounds	G. Triebig, A. Ihrig, M. Bader	2001
Mercury, organic mercury compounds	K.H. Schaller	2003
Methyl bromide	E. Hallier, M. Müller	2002
Phenol, Addendum	J. Lewalter, G. Leng	2003
Tetrachloroethene	K.H. Schaller, H.M. Bolt	1997 and 2001
Tetrachloromethane, Addendum	H.M. Bolt	2003
Tetrahydrofuran, Addendum	J. Lewalter, G. Leng	2000
Trichloroethene, Addendum	H.M. Bolt	2001

Working Group "Aufstellung von Grenzwerten in biologischem Material" (Setting of Threshold Limit Values in Biological Material)

Head: Prof. Dr. med. H. Drexler
        Institut und Poliklinik für Arbeits-, Sozial-
        und Umweltmedizin der
        Universität Erlangen-Nürnberg
        Schillerstr. 25/29
        D-91054 Erlangen

Members and guests of the Working Group:

Prof. Dr. rer. nat. J. Angerer
Institut und Poliklinik für Arbeits-, Sozial-
und Umweltmedizin der
Universität Erlangen-Nürnberg
Schillerstr. 25/29
D-91054 Erlangen

Prof. Dr. med. Dr. rer. nat. H.M. Bolt
Institut für Arbeitsphysiologie
Ardeystr. 67
D-44139 Dortmund

Prof. Dr. G. Csanády
GSF, Institut für Toxikologie
Ingolstädter Landstr. 1
D-85764 Neuherberg

Prof. Dr. med. H. Greim
Technische Universität München
Institut für Toxikologie
Hohenbachernstr. 15 – 17
D-85354 Freising-Weihenstephan

Prof. Dr. med. E. Hallier
Institut für Arbeits- und Sozialmedizin
Zentrum Umwelt- und Arbeitsmedizin
Georg-August-Universität
Waldweg 37
D-37073 Göttingen

Prof. Dr. rer. nat. H. Kappus
Karlsbergweg 8
D-87665 Mauerstetten

PD Dr. Dr. Udo Knecht
Institut und Poliklinik für
Arbeits- und Sozialmedizin
der Universität Giessen
Aulweg 129/III
D-35392 Giessen

PD Dr. G. Leng
Gesundheitsschutz und Arbeitssicherheit
Institut für Biologisches Monitoring, L9
Bayer Industry Services
D-51368 Leverkusen

Dr. rer. nat. J. Lewalter
Ärztliche Abteilung, Arbeitsstoffanalytik
Bayer AG
D-51368 Leverkusen

Prof. Dr. med. A.W. Rettenmeier
Institut für Hygiene und Arbeitsmedizin
Universitätsklinikum Essen
Hufelandstr. 55
D-45122 Essen

Prof. Dr. med. H.W. Rüdiger
Universität Wien
Abteilung Arbeitsmedizin
Med. Klinik IV
Währinger Gürtel 18-20
A-1090 Wien

Dipl.-Ing. K. H. Schaller
Institut und Poliklinik für Arbeits-, Sozial-
und Umweltmedizin der
Universität Erlangen-Nürnberg
Schillerstr. 25/29
D-91054 Erlangen

Prof. Dr. med. Dipl.-Chem. G. Triebig
Institut und Poliklinik für Arbeits- und Sozialmedizin
der Universität Heidelberg
Hospitalstr. 1
D-69115 Heidelberg

Prof. Dr. med. R. Wrbitzky
Institut und Poliklinik für Arbeitsmedizin
der Medizinischen Hochschule Hannover
Carl-Neuberg-Str. 1
D-30625 Hannover

Secretary: Institut und Poliklinik für Arbeits-, Sozial-
und Umweltmedizin der
Universität Erlangen-Nürnberg
Schillerstr. 25/29
D-91054 Erlangen

Index for Volumes 1-4

Acetic acid 2-butoxyethyl ester **3** 93
Acetic acid-2-ethoxyethyl ester **1** 71
Acetone 3 65–77
Acetylcholinesterase inhibitors 2 15–24
Acrylamide 3 243–248
Acrylic acid amide **3** 243–248
Acrylonitrile 2 183–192
Alkali chromates (Cr(VI)) 1 187–203
Alkyl phosphates **2** 15–24
Alkyl sulphates **2** 15–24
Aluminium 1 3–16
Aluminium hydroxide **1** 3–16
Aluminium oxide **1** 3–16
Aminobenzene **2** 25–34
4-Aminodiphenyl 2 193–201
p-Aminodiphenyl **2** 193–201
2-Aminonaphthalene **2** 213–221
2-Aniline 2 25–34
Antimony and its inorganic compounds
 4 141-149
Arsenic and inorganic arsenic
 compounds 4 171-178
Arsenic acid **4** 171-178
Arsenic pentoxide **4** 171-178
Arsenic trioxide 2 124–136
Arsenic trioxide **4** 171-178
Arsenous acids **4** 171-178
Benzene 3 217–231
Benzene-γ-hexachloride **2** 67–75
Beryllium and its inorganic compounds
 4 151-158
Blasting gelatin **3** 143–162
2-Bromo-2-chloro-1,1,1-trifluoroethane
 2 35-42
Bromomethane **3** 249–257
Bromomethane **4** 189-199
2-Butanone 2 43–50
2-Butoxyethanol 3 79–92
n-Butoxyethanol **3** 79–92
2-Butoxyethanol acetate **3** 93
2-Butoxyethyl acetate 3 93
Butyl Cellosolve® **3** 79–92
Butyl Cellosolve® acetate **3** 93
Butylene oxide **2** 99–105
O-Butyl ethylene glycol **3** 79–92
Butyl glycol **3** 79–92
Butyl glycol acetate **3** 93

***p-tert*-Butyl phenol 1** 17–21
Calcium arsenate **4** 171-178
Carbamates **2** 15–24
Carbinol **1** 99–114
Carbon disulphide 1 23–29
Carbon monoxide 1 31–35
Carbon tetrachloride **1** 153–161
Carbon tetrachloride **4** 117-118
Cellosolve® **1** 57–70
Cellosolve acetate® **1** 71
Chlorobenzene 1 37–44
Chloroethylene **2** 173–179
Cinnamene **3** 163–184
Cinnamol **3** 163–184
Cobalt 1 205–216
Cresols (all isomers) 4 179-188
m-Cresol **4** 179-188
o-Cresol **4** 179-188
p-Cresol **4** 179-188
m-Cresylic acid **4** 179-188
o-Cresylic acid **4** 179-188
p-Cresylic acid **4** 179-188
Cyclohexane 4 13-22
1,4-DCB **3** 95–104
p-DCB **3** 95–104
Diamide **2** 153–159
p,p'-Diaminodiphenylmethane **2** 203–212
4,4'-Diaminodiphenylmethane
 2 203–212
1,4-Dichlorobenzene 3 95–104
1,4-Dichlorobenzene 4 159
p-Dichlorobenzene **3** 95–104
Dichloromethane 1 45–55
Dichloromethane 4 123-126
Diethylene oxide **2** 99–105
Dimethylbenzene **2** 107–119
***N,N*-Dimethylformamide 2** 51–57
Dimethylketone **3** 65–77
Dimethyl sulfate 3 233–239
DMF **2** 51–57
DMS **3** 233–239
E 605 **2** 83–88
EGDN **3** 105–121
Ethene **2** 137–143
Ethenylbenzene **3** 163–184
2-Ethoxyethanol 1 57–70
2-Ethoxyethyl acetate 1 71

Ethylene 2 137–143
Ethylene glycol dinitrate 3 105–121
Ethylene glycol monobutyl ether 3 79–92
Ethylene glycol monobutyl ether acetate 3 93
Ethylene glycol monoethyl ether 1 57–70
Ethylene glycol monoethyl ether acetate 1 71
Ethylene glycol n-butyl ether 3 79–92
Ethylene oxide 2 145–151
Ethyl glycol 1 57–70
Ethyl glycol acetate 1 71
Ethyl methyl ketone 2 43–50
Formic acid methyl ester 4 23-29
N-Formyldimethylamine 2 51–57
Glycerol nitric acid triester 3 143–162
Glyceryl trinitrate 3 143–162
Glycide nitrate 3 105–121
Glycol butyl ether 3 79–92
Glycol dinitrate 3 105–121
Glycol monobutyl ether acetate 3 93
Glycol monoethyl acetate 1 71
Halothane 2 35–42
γ-HCH 2 67–75
Hexachlorobenzene 2 59–65
Hexachlorobenzene 4 31-37
γ-1,2,3,4,5,6-**Hexachlorocyclohexane** 2 67-75
Hexahydrobenzene 4 13-22
Hexamethylene 4 13-22
Hexanaphthene 4 13-22
n-**Hexane** 1 73–85
2-Hexanone 1 87
Hexone 1 115–122
Hydrazine 2 153–159
Hydrogen fluoride and inorganic fluorine compounds (fluorides) 1 89–97
1-Hydroxy-2-methylbenzene 4 179–188
1-Hydroxy-3-methylbenzene 4 179–188
1-Hydroxy-4-methylbenzenel 4 179-188
1-Hydroxy-4-$tert$-butylbenzene 1 17–21
Hydroxytoluene 4 179-188
2-Hydroxytoluene 4 179-188
3-Hydroxytoluene 4 179-188
4-Hydroxytoluene 4 179-188
Isobutyl methyl ketone 1 115–122
Isopropanol 1 129–138
Isopropyl alcohol 1 129–138
β Ketopropane 3 65–77
Lead and its compounds 4 39-87
Lead arsenate 4 171-188
Lead tetraethyl 2 89–98
Lead tetramethyl 3 185–187
Lindane 2 67–75
Manganese and its inorganic compounds 4 89-115
MBK 1 87
MDA 2 203–212
MEK 2 43–50
Mercury, metallic mercury and inorganic mercury compounds 3 123–142
Mercury, organic mercury compounds 4 161-168
Methanol 1 99–114
Methyl alcohol 1 99–114
Methylbenzene 3 189-204
Methyl bromide 3 249–257
Methyl bromide 4 189-199
Methyl-n-butyl-ketone 1 87
Methyl chloroform 1 163–170
Methylene chloride 1 45–55
Methylene chloride 4 123-126
4,4-Methylenedianiline 2 203–212
Methyl ethyl ketone 2 43–50
Methyl formate 4 23-29
Methyl isobutyl ketone 1 155–122
Methyl methanoate 4 23-29
4-Methylpentan-2-one 1 155–122
Methylphenol 4 179-188
2-Methylphenol 4 179-188
3-Methylphenol 4 179-188
4-Methylphenol 4 179-188
Methyl sulfate 3 233–239
MIBK 1 155–122
Monobromethane 4 189-199
Monobromomethane 3 249–257
Monobutyl glycol ether 3 79–92
Monochlorobenzene 1 37–44
2-Naphthylamine 2 213–221
β Naphthylamine 2 213–221
Nickel / sparingly soluble nickel compounds 2 161-172
Nitrobenzene 2 77–82
Nitroglycerin 3 143–162
Nitroglycerol 3 143–162
Nitroglycide 3 105–121
Nitrostigmine 2 83–88
Organic phosphates 2 15–24
Organic sulphates 2 15–24
3-Oxa-1-heptanol 3 79–92
Oxolane 2 99–105
Paraphos 2 83–88
Parathion 2 83–88
PCP 1 217–235

Pentachlorophenol 1 217–235
Perchlorobenzene 2 59–65
Perchloroethylene 1 139–151
Perchloroethylene 4 127-129
Phenol 1 123–128
Phenol 4 201-205
Phenylamine 2 25–34
Phenylethylene 3 163–184
Phenylmethane 3 189–204
1,2,3-Propanetriol trinitrate 3 143–162
2-Propanol 1 129–138
2-Propanone 3 65–77
2-Propenamide 3 243–248
Propenenitrile 2 183–192
Sodium arsenite 4 171-178
Stibium 4 141-149
Styrene 3 163–184
Styrol 3 163–184
Styrolene 3 163–184
Sulfuric acid dimethyl ester 3 233–239
Sulphonates 2 15–24
Tetrachloroethene 1 139–151
Tetrachloroethene 4 127-129
Tetrachloromethane 1 153–161
Tetrachloromethane 4 117-118
Tetraethyllead 2 89–98
Tetraethylplumbane 2 89–98
Tetrahydrofuran 2 99–105
Tetrahydrofuran 4 119-120
Tetramethylene oxide 2 99–105
Tetramethyllead 3 185–187
Tetramethylplumbane 3 185–187
Toluene 3 189–204
Toluol 3 189–204
1,1,1-Trichloroethane 1 163–170
Trichloroethene 1 171–183
Trichloroethene 4 131-137
Trichloroethylene 1 171–183
Trichloroethylene 4 131-137
Trinitroglycerol 3 143–162
Vanadic anhydride 3 205–214
Vanadium(V) oxide 3 205–214
Vanadium pentoxide 3 205–214
VCM 2 173–179
Vinylbenzene 3 163–184
Vinyl chloride 2 173–179
Vinyl cyanide 2 183–192
Xylenes 2 107–119